Façade Design - Challenges and Future Perspective

Edited by Chiara Bedon, Marcin Kozlowski and Mislav Stepinac

Published in London, United Kingdom

Façade Design - Challenges and Future Perspective
http://dx.doi.org/10.5772/intechopen.111334
Edited by Chiara Bedon, Marcin Kozlowski and Mislav Stepinac

Contributors
Poonam Singh Singh Kharwar, Devesh Kumar, Abhishek Kumar, Abhinav Kumar, Lucio Blandini, Christina Eisenbarth, Walter Haase, Moon-Young Jeong, Michael Voigt, Daniel Roth, Arina Cazan, Maria Matheou, Tianyi Chen, Chye Kiang Heng, Shin Woei Leow, Silvia Brunoro, Valentina Frighi, Mubarak Reme Ibrahim, Alcinia Zita Sampaio, Augusto Gomes, Inês Domingos, Chiara Bedon, Mohammad Momeni

First published in London, United Kingdom, 2024 by IntechOpen
IntechOpen is the global imprint of INTECHOPEN LIMITED, registered in England and Wales, registration number: 11086078, 5 Princes Gate Court, London, SW7 2QJ, United Kingdom
Printed in Croatia

British Library Cataloguing-in-Publication Data
A catalogue record for this book is available from the British Library

Additional hard and PDF copies can be obtained from orders@intechopen.com

Façade Design - Challenges and Future Perspective
Edited by Chiara Bedon, Marcin Kozlowski and Mislav Stepinac
p. cm.

This title is part of the Civil Engineering Book Series, Volume 4
Topic: Construction Engineering
Series Editor: Assed Haddad
Topic Editor: Amin Akhnoukh

Print ISBN 978-0-85014-186-3
Online ISBN 978-0-85014-187-0
eBook (PDF) ISBN 978-0-85014-188-7
ISSN 3029-0287

For EU product safety concerns:
IN TECH d.o.o., Prolaz Marije Krucifikse Kozulić 3, 51000 Rijeka, Croatia,
info@intechopen.com or visit our website at intechopen.com.

IntechOpen Book Series

Civil Engineering

Volume 4

Aims and Scope of the Series

Civil engineering is a traditional field of engineering from which most other branches of engineering have evolved. It comprises traditional sub-areas like transportation, structures, construction, geotechnics, water resources, and building materials. It also encompasses sustainability, risk, environment, and other concepts at its core. Historically, developments in civil engineering included traditional aspects of architecture and urban planning as well as practical applications from the construction industry. Most recently, many elements evolved from other fields of knowledge and topics like simulation, optimization, and decision science have been researched and applied to increase and evolve concepts and applications in this field. Civil engineering has evolved in the last years due to the demands of society in terms of the quality of its products, modern applications, official requirements, and cost and schedule restrictions. This series addresses real-life problems and applications of civil engineering and presents recent, cutting-edge research as well as traditional knowledge along with real-world examples of developments in the field.

Meet the Series Editor

Professor Assed N. Haddad is a Civil Engineer with a degree from the Federal University of Rio de Janeiro (UFRJ) earned in 1986, as well as a Juris Doctor degree from the Fluminense University Center earned in 1993, and a Master's degree in Civil Engineering from the Fluminense Federal University (UFF) obtained in 1992. He completed his Ph.D. in Production Engineering from COPPE / Federal University of Rio de Janeiro in 1996. Professor Haddad's academic pursuits have taken him to postdoctoral stays at the University of Florida, USA in 2006; at the Universitat Politècnica de Catalunya, Spain in 2010; and at the University of New South Wales Sydney, Australia in 2019. Currently, he serves as a Full Professor at the Federal University of Rio de Janeiro. He has held visiting professorships at various institutions including the University of Florida, Universitat Politècnica de Catalunya, Universitat Rovira i Virgili, and Western Sydney University. His research expertise encompasses Civil, Environmental, and Production Engineering, with a primary focus on the following topics: Construction Engineering and Management, Risk Management, and Life Cycle Assessment. He has been the recipient of research grants from the State of Rio de Janeiro, Brazil: CNE FAPERJ from 2019 to 2022 and from 2023 to 2025. Additionally, his research grants obtained from the Brazilian Government CNP since 2012 last to this date. Professor Haddad has been involved in several academic endeavors, being the Guest Editor of the International Journal of Construction Management; MDPI's Sustainability, Energies, and Infrastructures; Associate Editor at Frontiers in Built Environment / Sustainable Design and Construction; Guest Editor at Frontiers in Built Environment / Construction Management; and Academic Editor of the Journal of Engineering, Civil Engineering Section of Hindawi. He is currently a Professor of the Environmental Engineering Program at UFRJ and the Civil Engineering Program at UFF.

Meet the Volume Editors

Chiara Bedon is an Associate Professor of Structural Engineering in the Department of Engineering and Architecture, University of Trieste, Italy, where she teaches Structural Analysis, Steel Structures, and Structural Design and Rehabilitation. She received a Ph.D. in 2012 and has been involved in international research projects and networks since 2009. She is an expert member, national delegate, task leader, and advisor at organizations such as the Joint Research Centre – European Reference Network for Critical Infrastructure Protection (JRC-ERNCIP), North Atlantic Treaty Organization – Science for Peace and Security (NATO-SPS), European Cooperation in Science and Technology (EU-COST), and others. She is an editorial board member for several journals. She is a Top Italian Scientist (TIS) and Top Italian Women Scientist (TIWS) in civil engineering. She is a former member of the EU-COST Action TU1403 – "Adaptive Facades Network," and a core group member for the ongoing EU-COST Action CA20139 – "Holistic design of taller timber buildings – HELEN." She is a principal investigator for projects funded by the European Union, including the recently granted FIS2021-MUR "HOPgLAz" project.

Marcin Kozłowski is an expert in the structural use of glass in construction, the safety of glass structures, and numerical modeling of complex engineering issues, including the brittle fracture of glass and the simulation of both static and dynamic loads. Since 2020, he has been an associate professor in the Department of Structural Engineering, Silesian University of Technology, Gliwice, Poland. Professor Kozłowski has served as the project manager in two international research projects and several others with national funding. He is a national delegate to the European Committee for Standardization CEN/TC 250/SC 11 "Structural Glass" and serves as the chairman of the technical committee for the design of glass structures in the Polish Standardization Committee. He actively contributed to several research networks and projects, including the European Cooperation in Science and Technology (EU-COST) Action TU1403, "Adaptive Facades Network" and the MINIATURE 5 competition financed by the National Science Center (NCN) with a research proposal entitled "Analysis of the internal pressure in 3D bent insulating glass".

Mislav Stepinac is an Assistant Professor of Structural Engineering in the Faculty of Civil Engineering, University of Zagreb, Croatia. He is an instructor for courses in Earthquake Engineering, Existing Masonry Structures, Precast Reinforced Concrete Structures, and Concrete and Masonry Structures. As the principal investigator for the "ARES" project on the assessment and rehabilitation of existing structures, he is involved in the development of contemporary methods for the structural analysis of masonry and timber constructions. He is a core group member of the ongoing European Cooperation in Science and Technology (EU-COST) Action CA20139 – "Holistic design of taller timber buildings – HELEN". He is also co-author of several journal and conference papers.

Contents

Preface

Façade design is a challenging task in which multidisciplinary issues and aspects should be optimally considered and addressed. The primary reason for this is that multiple objectives are required for these building systems, from several points of view. This is especially the case for building façades that are expected to act as physical barriers and protect the occupants from extreme occurrences, such as seismic events, impacts, or fire. The use of special monitoring and technologies could be required for the residual capacity assessment of existing façades. Special attention and major efforts are also required for the detection and application of new technologies in the generation of modern, adaptive façade systems and building envelopes.

In this regard, to support the assessment and refinement of new strategies, technologies, trends, and challenges of façade design, this book presents a selection of research contributions from international scholars and façade experts, in which the aforementioned aspects are investigated by means of experimental methods, finite element modelling strategies, application of new technologies, and development of innovative design concepts. This book is designed for researchers and scholars as well as professionals in the field.

The selected book chapters are the results of recent research efforts of several authors as well as the support and input from a team of experts on the topic. We gratefully acknowledge the authors and reviewers for sharing their time and experience.

As editor of this book, I would like to thank my co-editors Prof. Marcin Kozłowski (Silesian University of Technology, Poland) and Asst. Prof. Mislav Stepinac (University of Zagreb, Croatia), for their fruitful scientific collaboration and for our long-term networking in the fields of structural engineering, dynamics, structural glass, and building façades.

Chiara Bedon
Associate Professor,
Department of Engineering and Architecture,
University of Trieste,
Trieste, Italy

Marcin Kozlowski
Associate Professor,
Department of Structural Engineering,
Silesian University of Technology,
Gliwice, Poland

Mislav Stepinac
Assistant Professor,
Faculty of Civil Engineering,
Structural Engineering,
University of Zagreb,
Zagreb, Croatia

Section 1

Experimental Numerical and Monitoring Tools

Chapter 1

Reimagining Building Façades: The Prefabricated Unitized BIPV Walls (PUBW) for High-Rises

Tianyi Chen, Chye Kiang Heng and Shin Woei Leow

Abstract

In urban settings, building-integrated photovoltaics (BIPV) on façades prove more effective than rooftop installations, especially for tall structures with limited roof area. Yet, the absence of ready-to-use BIPV solutions restricts their broader use. This research presents a prefabricated unitized BIPV wall system, using light gauge steel structure prefabrication. The innovative BIPV system boasts a multifunctional, modular design, ensuring quick installation and meeting airtightness standards. The design process encompasses cross-sectional design, PV mounting, 3D modeling, and full-scale mock-up demonstrations in Singapore. Remarkably, the prefabricated units are preassembled and pre-wired, allowing three non-specialized workers to install from inside buildings, eliminating the need for scaffolding. The study offers insights into the new BIPV system's advantages, identifies its constraints, and suggests avenues for future enhancement.

Keywords: building-integrated photovoltaics, BIPV, solar energy, prefabricated construction, opaque PV, photovoltaics implementation

1. Introduction

Over the past decade, worldwide greenhouse gas emissions have surged, hitting unprecedented levels in recorded history [1]. Nations aligned with the Paris Agreement are committed to slashing their carbon outputs. As a result, they are formulating energy blueprints to shift their energy sources and diminish dependency on fossil fuels.

Unlike building-attached photovoltaics (BAPV) that need extra structures for installation, BIPV systems integrate photovoltaic modules directly into the building's façade, serving both as an envelope and a power generator. With silicon costs dropping, PV module prices are now on par with common construction materials like marble and aluminum cladding [2]. While rooftops offer solar exposure, their space is limited by infrastructure equipment. In contrast, building façades offer ample space for PV integration, making them ideal for BIPV.

In urbanized cities, tall buildings dominate the skyline. Their design influences material choices, demanding innovative construction and safety solutions. Currently, the main issue with tall structures and BIPV façades is accessibility. The installation and construction require workers to use scaffolding or elevating platforms for PV module installations, often preceded by fixing aluminum cassettes. This process is not only tedious but resembles the construction approach of BAPV, emphasizing the challenges with time and costs previously noted [3]. Elevated BIPV work presents safety concerns, with PV components vulnerable to adverse weather, impacting performance and safety [4]. Some prefabricated curtain wall systems with PV technology offer installation from inside but compromise on power generation or integration areas.

Prefabricated construction is a method wherein building components are produced and assembled in an offsite factory before being transported to the construction site for erection [5]. This technology brings several advantages [6]. Primarily, it can expedite the process of on-site installation. Additionally, it allows for stringent quality control of construction in the controlled environment of the offsite factory, thus promoting material efficiency. Lastly, it mitigates the risk associated with laborers working in hazardous environments.

Presently, several studies are concentrating on the intersection of the photovoltaic industry with the prefabrication construction sector. RICS [7] proposed the concept of a prefabricated BIPV business model to reduce costs in industry development. Large prefabrication construction firms can establish dedicated PV departments, thereby eliminating the need for end-users to deal with contracts and maintenance of the PV system in their residences [8]. This arrangement also simplifies the process of accessing renewable energy subsidies. Further, Longas et al. [9] proposed the concept and prototype of a PV system featuring a laminated timber wood structure using prefabricated construction. Valckenborg [10] incorporated PV thin film within an aluminum folded structure to expedite construction. However, these current studies fall short of satisfying the requirements of high-density urban development when integrating with PV systems, as is the case in Asian countries like Singapore.

This study introduces a new design for a fully prefabricated BIPV wall suitable for tall structures, streamlining PV installation, and wall structuring without exterior scaffolding. The outcome is the prefabricated unitized BIPV wall (PUBW). This multi-layered, opaque BIPV wall minimizes on-site height-related risks, ensures efficient electricity generation, faster construction, cost savings, and protects PV components from prolonged adverse weather during building.

2. Various BIPV building integrations

The various types of BIPV building integrations introduced in the previous section can be mainly divided into two categories: roof system integration and façade system integration. The following illustrates the concept details of each type of building integration and the advantages and disadvantages when integrated with the PV panel.

2.1 Roof PV systems

Skylights use glazes to cover a part or entirety of a roof or atrium [11]. Skylights allow people to enjoy sunlight in an indoor environment. PV modules can replace glass to form semi-transparent or opaque parts in the skylight. By changing the density of the silicon crystal material on the glass, the indoor lighting and shading effects can be adjusted.

Cold roofs comprise an external cladding material, an air cavity, and a load-bearing substructure [12]. The ventilated air cavity dissipates the temperature of the outer cladding material. It is also known as a discontinuous roof because the cladding material usually forms a watertight layer of tiles, slate, sheet metal, etc. PV can replace the cladding material and mimic its color and texture through external coatings, making it "invisible". This type of PV integration is widely used especially where the roof receives sufficient solar radiation and has a certain angle of inclination.

A shed roof is the external extension space of a building with climate protection functions, such as shelter from sun and rain. PV replaces the cladding or glazing material of the canopy and does not require the same thermal performance considerations as skylights.

2.2 Façade PV systems

A rainscreen wall comprises cladding, an air cavity and a substructure bearing the load. This type of wall uses a ventilated air cavity to dissipate solar heat; hence, it is also called a cold wall [13]. PV modules replace the external cladding material and increase the efficiency of electricity production, particularly silicon PV, owing to rear ventilation. Slits could be present between the PV cladding and other cladding materials; hence, no significant pressure difference occurs between the cavity and the exterior. The water that penetrates inside the cavity eventually drains out of the cavity through evaporation and natural outflow. The substructure bearing load is protected against water.

A curtain wall is a strong representation of the industrialization of the building façade, mainly comprising an extruded aluminum frame and filled glass or metal panels, which can be modularly manufactured and pre-assembled, such as unitized curtain wall, or assembled on site, such as stick curtain wall [14]. Curtain walls need to meet the function of resistance of horizontal load and transfer to the main structure of the building, such as wind force and earthquake, as well as airtightness, heat insulation, sound proofing, etc. They are self-standing structures and bear no dead load weights from the building, hanging from the edge of the floor slab by bracket components. PV usually takes the place of filled clear glass (vision window) or metal plate/opaque glass (spandrel).

A BIPV double skin comprises two layers of glass. The outer skin is mainly intended for lighting purposes, while the inner glass has an insulating effect. The distance between the two layers of glass varies from about 20 cm to 2 m, acting as heat insulation, thermal insulation, sound insulation, etc. The ventilation method (e.g., mechanical, natural or hybrid) and opening and closing time control of the air cavity can be adjusted depending on the climate conditions of the building for energy saving. PV replaces part or the entirety of the outer skin of the glass.

BIPV shading devices are generally placed on the external surface of a building façade to control the intensity of light and heat entering indoors [15]. Various configurations and functional options are available, such as providing shade from top sunlight, the east morning sun or the west sun in the evening. Thus, the power generation function of PV and shading can well complement each other.

3. Prefabricated construction technologies

There are various construction materials that can be utilized for prefabricated construction, such as concrete, laminated timber wood, steel, and light gauge steel. This research incorporates a light gauge steel structure as the principal structural system due to its recyclable nature, which makes it apt for application in high-density

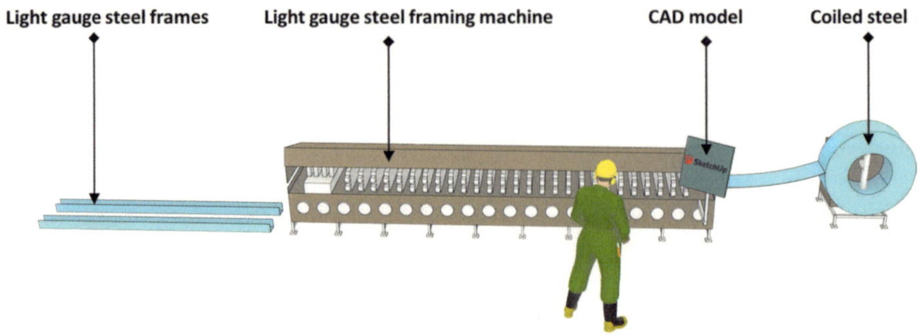

Figure 1.
From coiled steel to the creation of light gauge steel frames, the manufacturing process is automated in the factory, translating CAD models directly into finished products.

urban areas like Singapore. **Figure 1** illustrates the automated fabrication process of light gauge steel frame components in the factory setting, utilizing data imported from the digital model.

In Singapore, prefabricated prefinished volumetric construction (PPVC) leverages a light gauge steel structure as its primary architectural element to enable the creation of high-level prefabricated buildings. The low embodied carbon of light gauge steel structures confers environmental advantages, with the added benefit of recyclability. For the design at hand, a C-shaped channel with a 2 mm thickness is adopted as studs. The study positions the load-bearing studs at intervals of 1000 mm. To counteract seismic and lateral wind load forces, as well as to sustain the weight of photovoltaic (PV) panels, three potential systems—X-bracing, sheathing panel, or a combination of both—are proposed as a means to withstand forces in the horizontal plane [16]. In this research, a blend of X-bracing and sheathing panel is utilized. Beyond their use in building façades, light gauge steel structures demonstrate high versatility as they can also be incorporated into roofs and partition walls.

Utilizing computer-aided design and automation, the C-channel structure can be tailored for specific lengths and hole placements, optimizing infrastructure embedding and PV panel holding, which reduces costs and time. For projects with substantial PV areas, one can either customize PV sizes or use automated tools to modify the light gauge steel structure to fit the PV dimensions and design. Being lightweight, these structures can be preassembled off-site and conveniently transported for on-site installation.

4. The design of prefabricated unitized BIPV wall (PUBW)

4.1 2D cross-sectional design

This research examines the environmental factors to identify the materials and sites associated with each layer of the BIPV façade, aiming to establish a standard 2D cross-sectional design. Such factors should be taken into account for PUBW. **Figure 2** presents the load-determining factors under consideration, which influence the selection of materials to be implemented in the construction system.

The strategy for weather control (including barriers against water, air, and vapor) aims to ensure the wall's water and air seal. PV, serving as the rain screen system's

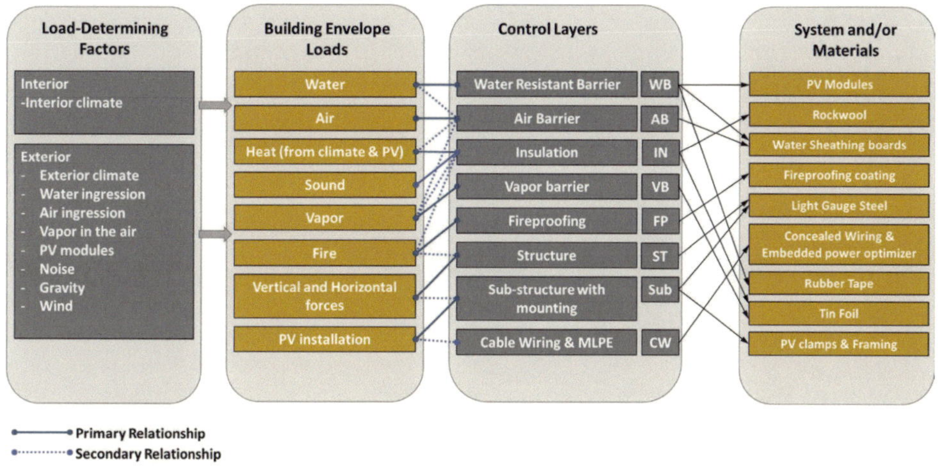

Figure 2.
Analysis of the functional layers in PUBW components.

cladding, is naturally waterproof, preventing most rainwater from coming into direct contact with the external wall and wires. Additionally, the seals between units must uphold weatherproofing standards.

Each PV module is equipped with a module-level power electronics (MLPE) to enhance power generation, especially when PV modules experience partial shading. Having pre-wired and pre-installed MLPE devices simplifies the installation for workers without electrical expertise.

Figure 3 depicts the standard horizontal and vertical cross-sections of PUBW. BIPV façade designs must meet core structural and weatherproofing standards to ensure a secure and cozy interior for inhabitants.

4.2 Mounting design

PV mounting systems greatly influence solar systems' esthetics and efficiency due to shading potential and load bearing. They fall into two categories: linear- and point-fixing systems [9]. Linear systems include mullion-transom and structural sealant systems. Mullion-transom systems have protruding elements causing potential for shadowing and debris accumulation. Structural sealant systems, without protrusions, use sealants for holding PVs and require extra precautions if above 8 m. Point-fixing systems are of three types: drilled spot, clamp fixing, and undercut anchor. Drilled spot systems demand specific PV types due to hole placements. Clamp fixing systems employ U-shaped clamps, while undercut anchor systems, reducing shading, need specialized glass types. This study's prototype employs the mullion–transom system for PV mounting.

4.3 Joint design

Joint design is crucial for functional continuity in independent prefabricated units [17]. It needs to ensure waterproofing, provide a "plug-and-play" assembly, and permit indoor manual installation without outdoor equipment like cranes. The PUBW's steel frame is consistently thin as it is non-structural. An "interlocking" design, like a

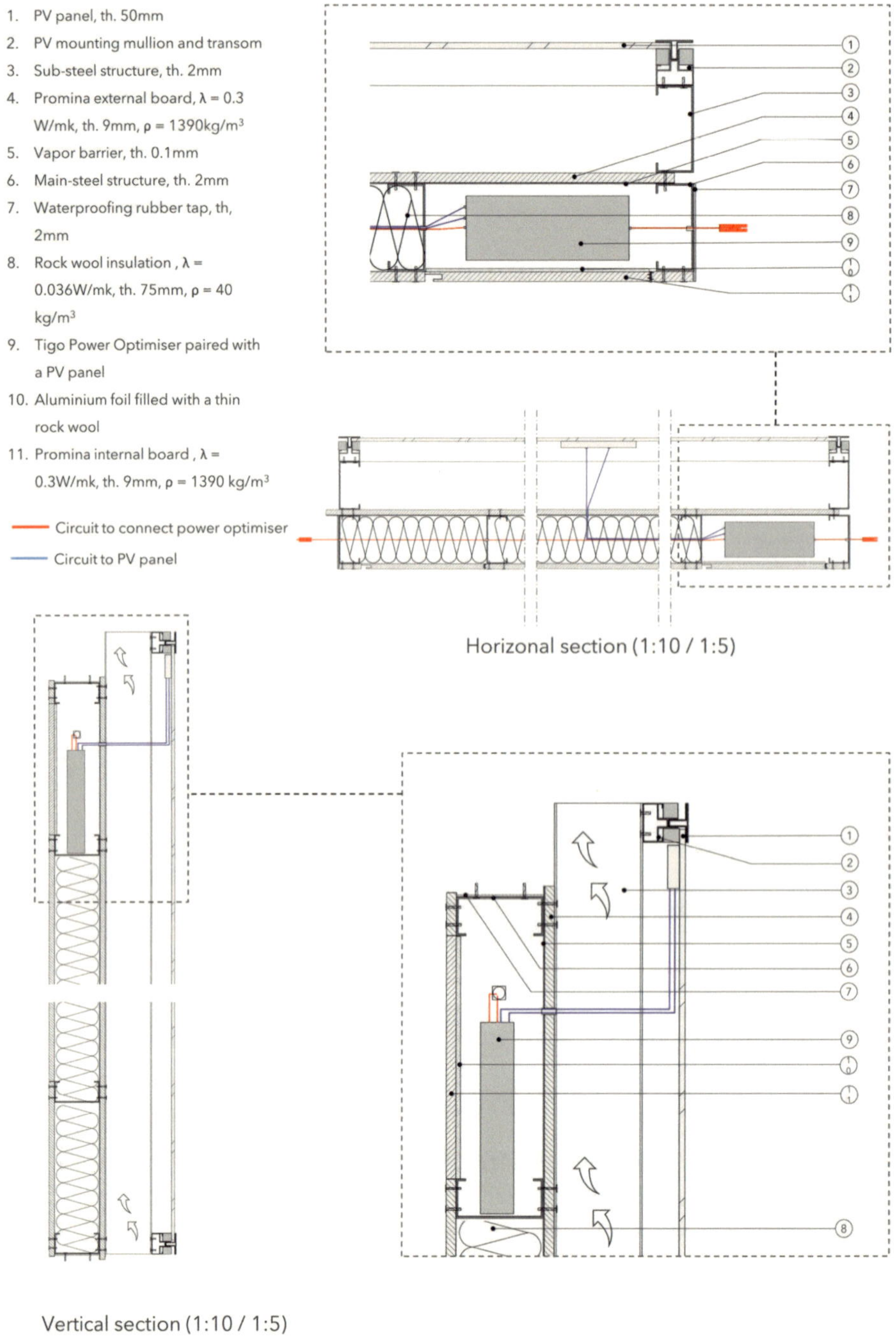

Figure 3.
2D horizontal cross-section of the preassembled unitized BIPV façade.

unitized curtain wall system, ensures seamless fitting and weatherproofing between adjacent units (**Figure 4**). This design uses male/female junctions for waterproofing and easy installation. Corners of the units allow for connections, making installations

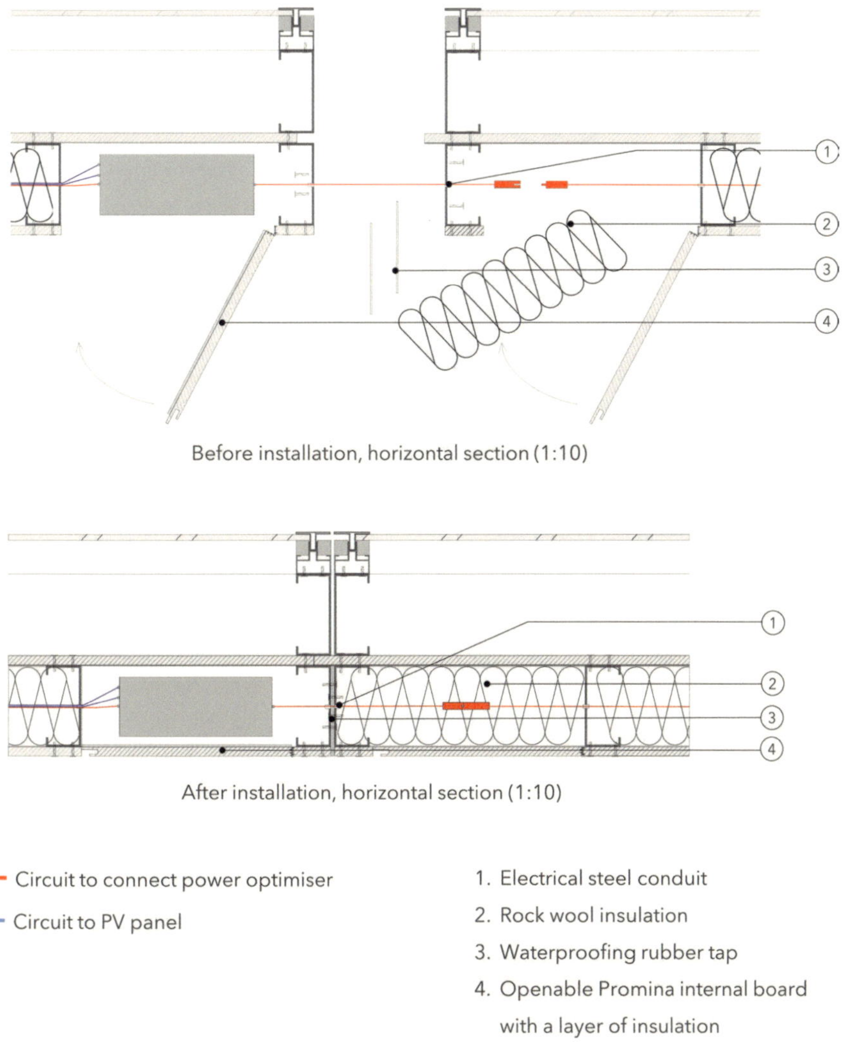

Figure 4.
Interlocking joints ensure a weatherproof and swift dry installation.

possible with just manual labor. Post-installation, equipment like MLPE devices fit into reserved corners, protected by rock wool insulation, ensuring no overheating. Accessible plasterboards ease future maintenance. After assembly, the interior wall gets painted for visual consistency. This system's design offers architects flexibility, as shown in combined sections with different unit types.

4.4 3D Modeling design

3D models of the PUBW units are created for initial system validation, which displays each layer's representation, clarifying design details (**Figure 5**). Moreover, **Figure 6** provides a glimpse into the off-site pre-assembly, aiding in worker training for fabrication and installation.

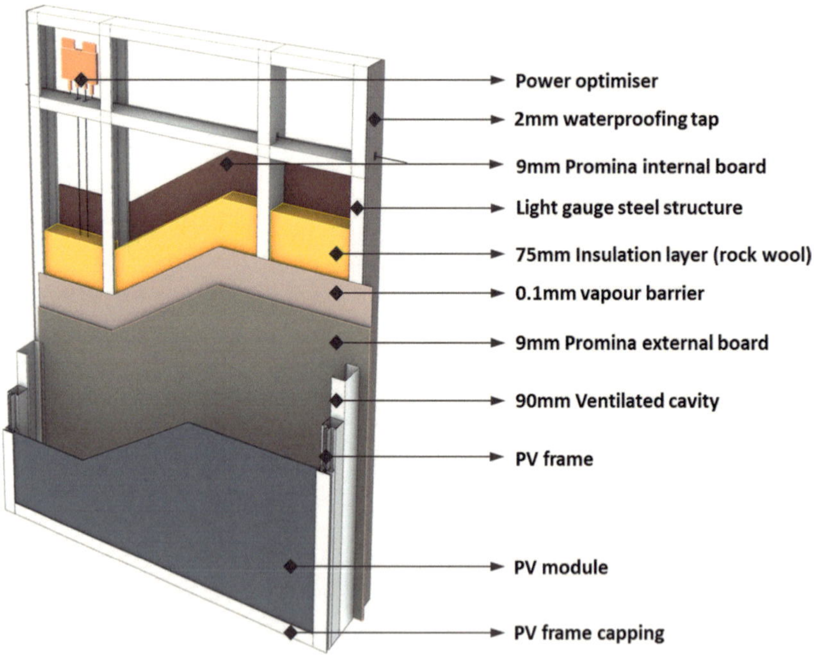

Figure 5.
3D model of the PUBW.

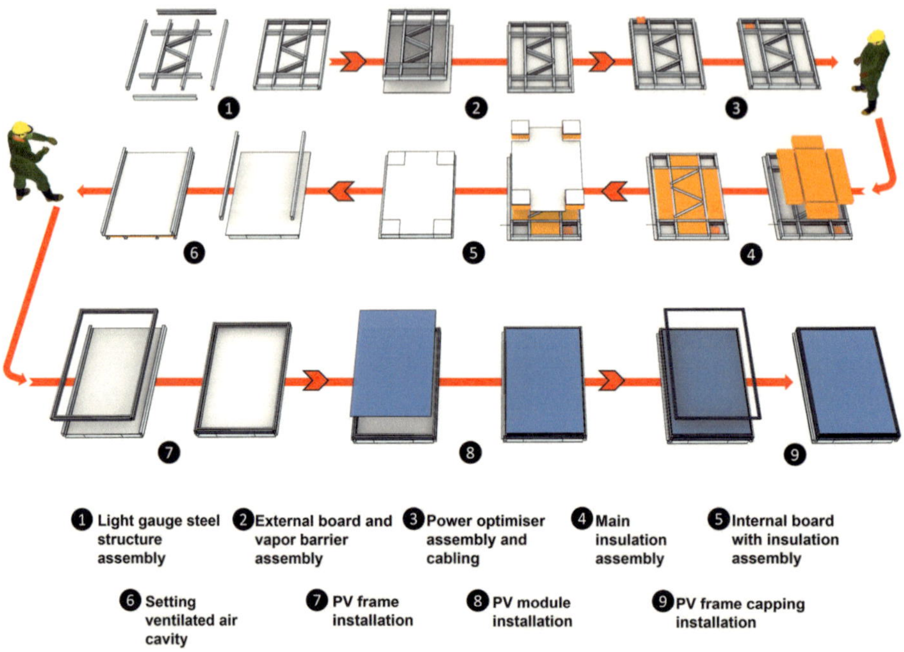

Figure 6.
Workers can be trained to understand the preassembly process through 3D models.

Figure 7.
The mock-up of the PUBW.

4.5 System mock-up

The "Modular Pod" prototype, made of light steel, showcased the PUBW on its eastern side (**Figure 7**). It was divided into two floors to replicate high-rise construction. Using standard building components and commercial PV modules, the prototype reflected actual dimensions. The installation was efficient, needing only three workers for assembly and PV wiring. After completion, the interior was painted for esthetics.

5. Conclusions

This research introduces a PUBW that combines architectural assembly and PV-powered energy, facilitating easy, scaffold-free installation on-site. Ideal for tall residential buildings with predominantly opaque façades on the west and east sides, this prefab façade shortens construction duration and simplifies BIPV system installation. Furthermore, it offers architects considerable design versatility and supports tailored mass production.

Technology can be adjusted to accommodate varying construction scenarios. If lifting or hoisting equipment is available on-site, the size of the PUBW can be scaled up to form a mega panel composed of several individual panels. Architects are free to design the material combination within the mega panel. For instance, they can integrate metal sheet panels with PV panels or modify the color of the panels. The larger size of the mega panel, compared to the standard size of PUBW, allows for faster and easier construction with the aid of hoisting machinery. This significantly increases the efficiency of on-site installation.

Furthermore, to enhance the efficacy of BIPV construction, it is of paramount importance to develop a comprehensive Building Information Modeling (BIM) digital library for PUBW or analogous BIPV prefabricated products. Digitized components within BIM can be viewed as a composite of multifaceted information, encompassing data on product logistics, associated costs, and photovoltaic performance metrics. By incorporating these digitized components into their BIM designs, architects can significantly economize on time, facilitate a more streamlined design procedure, and consequently accelerate the iterative process of design.

Acknowledgements

This work was conducted under a Solar Competitive Research Program grant from the National Research Foundation Singapore (NRF) through the Singapore Economic Development Board (EDB). The project "Cost-effective high-power density BIPV modules" (R-712-000-083-272) is implemented by the Solar Energy Research Institute of Singapore (SERIS) in collaboration with the Department of Architecture in the College of Design and Engineering (CDE) at the National University of Singapore (NUS). SERIS is a research institute at the National University of Singapore (NUS). SERIS is supported by the NUS, the National Research Foundation Singapore (NRF), the Energy Market Authority of Singapore (EMA) and the Singapore Economic Development Board (EDB).

Author details

Tianyi Chen[1*], Chye Kiang Heng[2] and Shin Woei Leow[1]

1 Solar Energy Research Institute of Singapore (SERIS), National University of Singapore, Singapore

2 Department of Architecture, National University of Singapore, Singapore

*Address all correspondence to: tianyi@nus.edu.sg

References

[1] Vadikkeettil Y et al. Plasma assisted decomposition and reforming of greenhouse gases: A review of current status and emerging trends. Renewable and Sustainable Energy Reviews. 2022;**161**:112343. DOI: 10.1016/j.rser.2022.112343

[2] SUPSI. Building Integrated Photovoltaics: A Practical Handbook for Solar Buildings' Stakeholders Status Report 2020. Switzerland: Swiss BIPV Competence Centre and Becquerel Institute; 2020 [Online]. Available from: https://solarchitecture.ch/wp-content/uploads/2021/02/BIPV_Status_Report.pdf [Accessed: July 31, 2023]

[3] Ahmadi MM et al. Net-zero exergoeconomic and exergoenvironmental building as new concepts for developing sustainable built environments. Energy Conversion and Management. 2021;**244**:114418. DOI: 10.1016/j.enconman.2021.114418

[4] Aghaei M et al. Review of degradation and failure phenomena in photovoltaic modules. Renewable and Sustainable Energy Reviews. 2022;**159**:112160. DOI: 10.1016/j.rser.2022.112160

[5] Luong DL, Nguyen QT, Pham AD, Truong QC, Duong MQ. Building a decision-making support framework for installing solar panels on vertical glazing Façades of the building based on the life cycle assessment and environmental benefit analysis. Energies. 2020;**13**(9):2376. DOI: 10.3390/en13092376

[6] Singapore Building Construction Authority, Design for Manufacturing and Assembly (DfMA). Singapore: Singapore Building Construction Authority; 2018 [Online]. Available from: https://www1.bca.gov.sg/docs/default-source/docs-corp-news-and-publications/publications/for-industry/ppvc_guidebook.pdf [Accessed: August 6, 2023]

[7] RICS. Cost Reduction and Deployment of Prefabricated Building Integrated Photovoltaics. Australia: The Royal Institution of Chartered Surveyors; 2019

[8] Yang RJ, Zou PXW. Building integrated photovoltaics (BIPV): Costs, benefits, risks, barriers and improvement strategy. International Journal of Construction Management. 2016;**16**(1):39-53. DOI: 10.1080/15623599.2015.1117709

[9] Longas AC, Lufkin S, Rey E. The construction of a full-scale demonstrator for BIPV architectural integration. In: Conference PLEA 2018 Smart and Healthy within the 2-degree Limit, Hong Kong. 2018

[10] Valckenborg RME. Zigzag structure IN FAÇADE optimizes PV yield while aesthetics are preserved. In: 32nd European Photovoltaic Solar Energy Conference and Exhibition : Proceedings of the International Conference Held in Munich, 20 June-24 June 2016, Germany. 2016. pp. 2481-2486

[11] Frontini F, Bonomo P, Moser D, Maturi L. Building integrated photovoltaic facades: Challenges, opportunities and innovations. In: Rethinking Building Skins. Netherlands: Elsevier; 2022. pp. 201-229. DOI: 10.1016/B978-0-12-822477-9.00012-7

[12] Ahmadi MM et al. Towards sustainable net-zero districts using the extended exergy accounting concept. Renewable Energy. 2022;**197**:747-764. DOI: 10.1016/j.renene.2022.07.142

[13] Day G, Gasparri E, Aitchison M. Knowledge-based Design in

Industrialised House Building: A case-study for prefabricated timber walls. In: Bianconi F, Filippucci M, editors. Lecture Notes in Civil EngineeringDigital Wood Design. Vol. 24. Cham: Springer International Publishing; 2019. pp. 989-1016. DOI: 10.1007/978-3-030-03676-8_40

[14] Gasparri E, Aitchison M. Unitised timber envelopes. A novel approach to the design of prefabricated mass timber envelopes for multi-storey buildings. Journal of Building Engineering. 2019;**26**:100898. DOI: 10.1016/j.jobe.2019.100898

[15] Yoon SD, Vuthy S, Choi HS. Design of Solar Modules for building Façades at educational facilities in Korea. Energies. 2021;**14**(9):2441. DOI: 10.3390/en14092441

[16] Di Lorenzo G, De Martino A. Earthquake response of cold-formed steel-based building systems: An overview of the current state of the art. Buildings. 2019;**9**(11):228. DOI: 10.3390/buildings9110228

[17] Orlowski K, Shanaka K, Mendis P. Manufacturing, Modeling, implementation and evaluation of a weatherproof seal for prefabricated construction. Buildings. 2018;**8**(9):120. DOI: 10.3390/buildings8090120

Chapter 2

Review on Glass Curtain Walls under Different Dynamic Mechanical Loads: Regulations, Experimental Methods and Numerical Tools

Mohammad Momeni and Chiara Bedon

Abstract

This chapter explores the behaviour and performance of glass curtain wall systems under various dynamic mechanical loads, including seismic, wind and impulsive loads. The classification of glass façade systems, comprising framed and frameless types, is first shortly discussed, along with their core components such as glass panels and frames. The challenges posed by glass material, including its vulnerability to impact, stress peaks and extreme loads, are acknowledged. The study further delves into various design standards and regulations for glass façade systems under dynamic loads, addressing seismic events and wind and impulsive loads and hence outlining parameters for assessment, performance criteria, and design considerations in use of glass curtain walls. Additionally, numerical methods are explored as effective tools for simulating and analysing the mechanical response of glass curtain walls under dynamic loads. The utility of these methods is showcased through a case study involving the Finite Element (FE) modelling of a glass curtain wall system exposed to a lateral in-plane load. The results of FE analysis are then compared with literature experimental results, which indicates its capacity to anticipate structural responses and even complex mechanisms under dynamic loads.

Keywords: glass curtain walls, dynamic loads, finite element models, experiments, regulations

1. Introduction

In modern structural and architectural design of buildings, glass curtain wall (CW) systems have emerged as a defining building feature that brings aesthetics as well as functionality for specific purposes. These non-structural systems consist of assemblies of glass panels supported by metal structures, which are connected to the main building through special connectors. Due to the relatively low tensile strength and brittle nature of glass as a load-bearing material in these systems, and with an increasing use of these systems in building designs, understanding their behaviour under various loads becomes important. This paper presents a review that explores the assessment of glass CW systems under different loads, including seismic, wind and impulsive loads. The first part of this paper looks into the fundamental principles of glass CW systems, distinguishing between framed and frameless configurations. The next parts delve into the realm of dynamic loads, encompassing seismic events, wind and impulsive loads such as blast and impact. The vulnerabilities and challenges posed by the glass material are then discussed, highlighting the need to address its relatively low tensile strength and susceptibility to brittle fracture. To this aim, an extensive review of available design standards for glass façade systems under dynamic loads as well as the classification of glass material based on their production methods is presented. The exploration of various design standards and codes underscores the collective efforts to ensure the resilience of these systems in the face of various dynamic forces. The utilisation of static and dynamic racking tests, wind tunnel experiments, shock tube tests and impact assessments elucidates the methodology behind ensuring the performance and safety of glass CW systems under dynamic loads. Moving beyond standards, the paper unfolds numerical methods that emerge as crucial tools for assessing structural responses under dynamic loads. Recognising the limitations and complexities of laboratory experiments, the significance of numerical methods, particularly finite element analysis, in comprehending the behaviour of glass CW systems under dynamic loading is also discussed. To validate the accuracy of numerical methods, an illustrative example of a finite element model is presented to evaluate the behaviour of a dry-glazed CW system under lateral in-plane loading.

2. Glass CW technology

A CW is a peripheral structure for buildings, which is composed – in most of cases - of metal supporting structures (aluminium or steel frame members) and plates (which can be composed of glass, aluminium, slate, ceramics, sandwich panels, etc.). When the panel is made of glass, as it is for CWs major functions of load-bearing capacity and architectural impact are expected. The classification of glass façade systems is based on their structural support, hence resulting in two main types: framed or frameless glass façade systems. Framed systems are typically designed using extruded aluminium components, although earlier versions used steel. On the other hand, frameless systems are restrained using bolted spider arms and steel supports, which serve as crucial architectural elements blending stability and aesthetic impact. The metal frames offer an efficient point support for glass panels but include various point-fixed joint options for truss and cable-supported systems. In these cases, typical systems comprise four key components: glass panels, bolted fixtures, glass support attachments (spider arms) and the main structural support frame [1]. Examples of framed and frameless glass façades are shown in **Figure 1**.

In this context, it is worth to remind that a glass CW represents a non-structural, exterior wall system that is used to clad buildings [2]. It is designed to separate the interior environment of a building from the external space while allowing light to enter and providing an important aesthetic appearance. Unlike more solid alternatives such as masonry, CWs are notably lightweight solutions of large use in modern

Figure 1.
Examples of glass façade; (a) framed glass façade, and (b) frameless glass façade.

constructions and derive their name from the resemblance to hanging curtains. Glass CW systems come then in two main forms: stick-built and unitised solutions, each with distinct structural attributes [3]. Stick-built systems involve on-site assembly of framing and glazing components, while unitised systems are typically pre-fabricated into CW units before installation. These constructional methods and technologies result in varying mullion-to-transom joints, where stick-built systems use aluminium shear blocks and unitised systems rely on direct screw connections. The unitised version also employs specific mullion profiles for easier on-site assembly and improved drift capacities under in-plane lateral loads, compared to stick-built systems. The choice between dry-glazed and structural sealant-glazed (SSG) configurations for glass panels is another parameter that further influences the mechanical performance of the assembled system [3]. Dry-glazed systems utilise rubber gaskets for compression, water resistance, and air infiltration prevention, while SSG systems are based on structural sealants to enhance water intrusion resistance and restrict the movement of glass panels. In addition, anchorage attachments play also a significant role, which is mostly influenced by factors like span lengths, temperature fluctuations, design loads and considerations for seismic or wind events and determining the size, shape and placement of attachment anchors. These attachments lead unavoidably to diverse responses during wind and seismic events, accommodating movements and rotations while maintaining flexibility. Overall, the interplay of these assembled components largely shapes and governs the complex behaviour of glass CW systems, underscoring the need for comprehensive engineering analysis, thoughtful design and construction considerations.

3. Glass material

The utilisation of glass in constructing envelopes has garnered substantial attention from researchers in recent decades [4]. However, it is important to remind that glass itself, despite its prevalent mechanical use, still poses challenges when overloaded, due to its relatively low tensile strength and brittle nature. This becomes particularly evident in the context of glazing windows and façades, which constitute delicate and breakable elements within a building structure. This vulnerability is especially pronounced when the design anticipates extreme loading conditions or the potential for such conditions arising over the structure's lifespan, as these glass envelopes serve as the primary barrier between the interior and exterior environments [5]. Finally, it is important to emphasise that glass components are the most vulnerable component in these systems, but an optimal structural design should pay attention for many other CW components [3]. With a focus on glass material, common types are classified based on their production methods into three categories [6]. AN glass material stands out for its cost-effectiveness due to its relatively straightforward production process. However, in terms of strength, it lags behind HS and FT glasses. Through the method of heating and gradual cooling used to transform AN glass, HS glass is formed, which is characterised by a certain surface compression within the glass panels. This results in a significant strength boost, compared to AN glass, and approximately doubles its strength. This arises from the harmonious distribution of thermal strains across the glass thickness during the heating and cooling phases, leading to a surface compressive stress raising up to about 30 MPa. On the other hand, elevating AN glass temperature to around 700°C and rapidly cooling it generates FT glass. In this case, the surface compression stress is notably high, exceeding 69 MPa as per ASTM C1048 standard [7]. This remarkable surface compression endows FT glass with a strength that is about 4 to 5 times greater than AN glass. Unlike AN and HS glasses, FT glass possesses stored elastic energy. Consequently, when broken, it fragments into numerous small, fine glass cubes (**Figure 2**). This unique behaviour is responsible for FT glass being commonly referred to as "safety glass" [8]. Worth noting is that under circumstances involving high-strain rate loads like explosions or impacts, FT glass, despite its safety characteristics, does hold the potential to break into larger pieces.

It should be noted that research studies and regulations emphasise the remarkable strength of glass material under high strain rate loads, with a dynamic increase factor of approximately 1.78, resulting in a compressive strength of 80 MPa for glass when subjected to impact or explosion [9, 10].

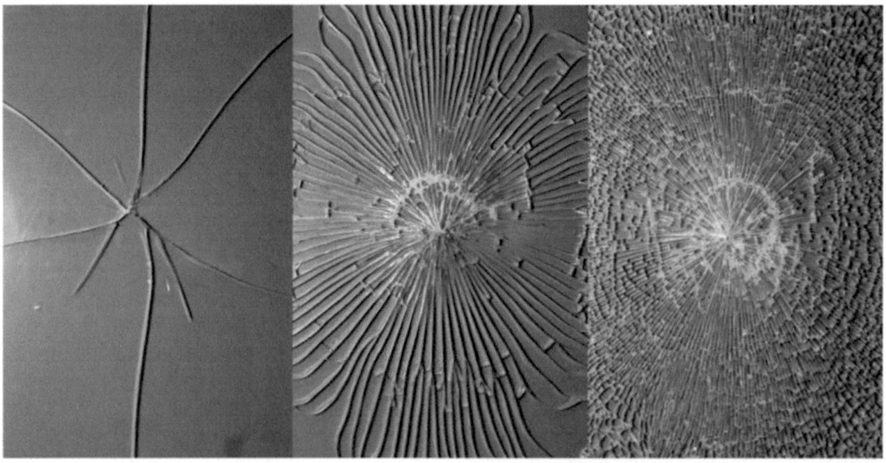

Figure 2.
Fracture pattern in AN (left), HS (middle), FT (right) glass material [8].

4. Selection of design standards and experimental protocols for glass façades under dynamic loads

Dynamic loads refer to forces that fluctuate in strength and direction over time. Such loads stem from sources like earthquakes, explosions, impacts, and sudden shifts in motion. They can induce extra stress on structures, materials, and components, necessitating their consideration during engineering and design to prevent fatigue, resonance, and other dynamic effects. In recent decades, significant events have caused substantial consequences for buildings and their envelopes. Notably, major earthquakes like the 1995 Northridge Earthquake prompted professionals to delve into improved glazing system design for enhanced safety, where the damage of glass façades was extensively observed, and more than 60% of the panes were broken [11]. Furthermore, recent explosion incidents have amplified the need for designing blast-resistant buildings, particularly those featuring CWs, to safeguard occupants from external explosions. Regrettably, many of these events have led to casualties, injuries and substantial financial losses, underscoring our built environment vulnerability (i.e. explosions in Tarragona, Spain 2020 [12], and Beirut, Lebanon in 2020 [13]). This underscores the necessity of considering blast loads when designing structures with CWs located near such facilities. Conversely, other loading types, such as wind and fire, should also be accounted for in glass CW design. It is worth noting that in areas with high wind velocities, wind governs the structure design, requiring the entire glass CW and its connections to withstand wind loads. Consequently, these components (i.e. glass CWs) typically face a multi-risk environment, subject to different types of loading as given above, and have often incurred significant damage, posing threats to life safety and incurring economic losses due to repair expenses and downtime.

4.1 Seismic events

During seismic events, the glass panels within the CW framing system experience in-plane displacement due to increasing story drift from the seismic forces. As stated in ASCE 7-16 [14], engineers are obligated to ensure that the relative seismic displacement (drift) of a considered glass CW component, D_{pI}, remains below the relative seismic displacement at which glass fallout from the CW, storefront, or partition occurs. This means that $\Delta_{fallout}$, as presented in Eq. 1, should respect the condition:

$$\Delta_{fallout} \geq max\left\{1.25D_{pI}, 13mm\right\} \quad (1)$$

It should be also noted that $D_{pI} = D_p I_e$, where D_p is the relative seismic displacement that the component must be designed to accommodate, and I_e is the importance factor (1.00, 1.25, 1.50 for increasing importance). However, there are some exceptions in ASCE 7-16 to describe states that do not comply with this requirement. In this regard, glass panels with sufficient gap from the frame, such that physical contact does not occur at the design drift do not need to satisty Eq. 1. Instead, the focus shifts to meeting the following criteria:

$$D_{clear} \geq 1.25D_{pI} \quad (2)$$

where D_{clear} signifies the relative seismic displacement at initial glass-to-frame contact. For rectangular glass panels, D_{clear} is determined by Eq. 3 as follows:

$$D_{clear} \geq 2c_1\left(1+\frac{h_p}{b_p}\frac{c_2}{c_1}\right) \tag{3}$$

where the rectangular glass panel height and width are denoted by h_p and b_p, respectively, and c_1 and c_2 represent the clearances between the frame and the vertical and horizontal edges of the glass, respectively. These parameters are shown in **Figure 3**.

As many other national and international standards for structural design, ASCE 7-16 uses "risk categories" to find an appropriate design wind velocity for determining the corresponding pressures and thus design structures and building components [14]. In this context, fully tempered monolithic glass in risk categories I, II and III and located within 3 m of a walking surface is exempt. Also, annealed or heat-strengthened laminated glass with an interlayer of at least 0.76 mm, mechanically captured in a wall system glazing pocket and secured to the frame by a wet-glazed elastomeric sealant perimeter bead (minimum 13 mm glass contact width), or an approved anchorage system, is not subjected to this requirement. It should be noted that different global codes present diverse performance criteria for CWs [16]. Eurocode 8 [17] offers permissible inter storey drift ratio values for damage control in non-structural elements under various conditions: 0.5%, 0.75% and 1% for brittle non-structural elements attached to the structure and for ductile non-structural elements or those integrated without obstructing structural deformation, respectively. New Zealand standard for structural design actions, NZS 1170.5 [18], dictates that a glass CW is considered to have failed when the relative displacement attains the larger of

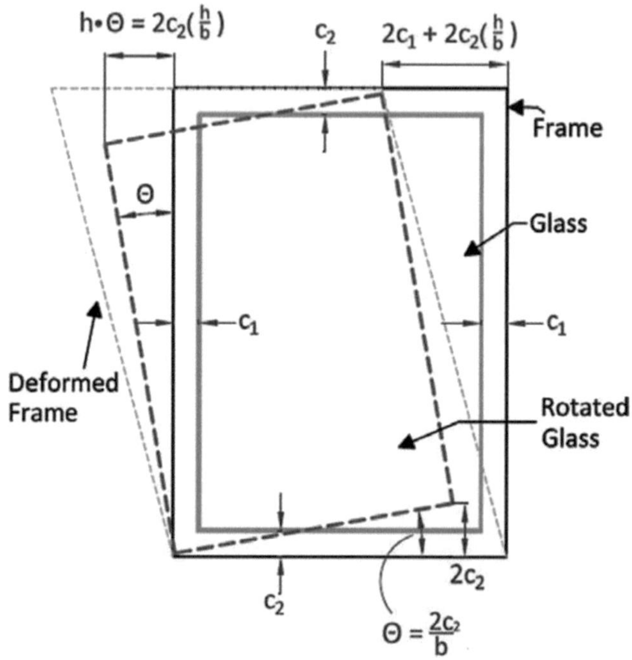

Figure 3.
Illustration of geometric parameters to determine drift at initial glass-to-fame contact [15].

two values: either 1/250 of the span or twice the width of the glass clearance. Here, the span represents the height of the story for the CWs attached to the building, while the glass clearance corresponds to the width of the silicone adhesive bar inserted on each side of the glass panel. In FEMA 273 [19], diverse configurations of glazing systems involve subframes attached to the main structure, either field-assembled or prefabricated. These systems are sensitive to deformations, and drift analysis is crucial to ensure compliance with performance levels. Failures, particularly in dry glazing, can result in shattering or detachment. Visual evaluation encompasses factors like glass support, mullion arrangement, sealants and connectors. Acceptance criteria revolve around force provisions and displacement for varying levels, with drift limits set at 0.02 and 0.01 for Life Safety and Immediate Occupancy performance levels, respectively. In FEMA P-58–2 [20–22], a fragility class is assigned to each category of glass CW, serving as a valuable point of reference for seismic design and calculations. In other words, FEMA P-58 employs repair costs instead of structural parameters to assess seismic performance. Fragility and repair cost functions are linked to vulnerable building components, showing the probability of surpassing damage thresholds at specific engineering demand parameter values. These functions, combined with peak structural responses, predict damage states and estimate economic losses. Other codes such as Chinese [23, 24], Canadian [25] and Japanese [26] establish varying limits tied to earthquake load intensity. In Chinese design codes [23, 24] for CWs, a distinct recommendation is presented. It stipulates that the in-plane peak drift of the CW should exceed three times the elastic deformation limit of the main structure. The Canadian code [25] assigns a value of 0.02 applicable to all structural types, which is a conservative approach across various building categories. It is understood that the glass panel loses its functionality immediately upon breakage. As per the Japanese code [26], glazed systems must be designed to adhere to inter storey drift ratio limits of 1% for severe earthquakes, 0.5% for moderate earthquakes, and 0.33% for low earthquakes. While the international codes' approaches mentioned above pertain to seismic requirements for glass CWs, the challenge persists in evaluating the CWs' capacity under various limit states. Generally, the codes emphasise the necessity of conducting experimental tests to address this concern. In the following, existing experimental tests to determine the lateral capacity of CW under seismic loading are discussed.

To evaluate the seismic performance of CW systems, various experimental protocols including shaking table test, in-plane racking test and so on can be found in the literature. These protocols serve as essential tools in understanding the dynamic behaviour of CW systems and contribute to more resilient and reliable structural designs. Shaking table testing stands as a foundational approach in assessing the seismic performance of CW systems. This technique involves subjecting scaled models or prototypes to simulated seismic motions, enabling insights into dynamic properties, system responses and failure modes under controlled conditions. Standardised codes such as AC156 [27] establish guidelines for seismic qualification tests of non-structural components and systems, which are adaptable for glass CW systems meeting specific criteria (i.e. systems with fundamental frequencies greater than 1.3 Hz). Similarly, FEMA 461 [28] guides the fragility evaluation of systems sensitive to dynamic motion. Despite not being explicitly tailored for shaking table testing of CWs affixed to buildings with multiple attachment points at neighbouring floor levels, this protocol nonetheless offers valuable insights into the testing of such CW systems. In-plane racking test serves as a pivotal method to assess the drift capacity of glass CW systems under dynamic loads. Standards like AAMA 501.4 [29] and AAMA 501.6 [30] focus on the in-plane draft capacity of framed glass façade systems as shown in **Figure 4**, offering methods for both static and dynamic testing. In AAMA 501.4, a horizontal static monotonic displacement is applied to the glass CW specimen up to a designated displacement. This facilitates the assessment of serviceability limit states, including factors like air infiltration, water penetration and structural integrity.

The second procedure, AAMA 501.6, takes a dynamic approach by subjecting the glass CW system to cyclic horizontal forces. This dynamic assessment, conducted with incremental concatenated sine waves following a crescendo pattern as depicted in **Figure 5**, mimics the stress dynamics experienced during seismic events. According to the protocol (more details about the incremental step loads can be found in [15]), the crescendo test should continue until one of these conditions is met: (1) a glass piece with an area of at least 645 mm^2 falls out, (2) the drift index (defined as the lateral displacement at the top of the glass panel divided by the glass panel height) is equal to or exceeds 0.1 (equivalent to 10%) or (3) a maximum racking displacement of

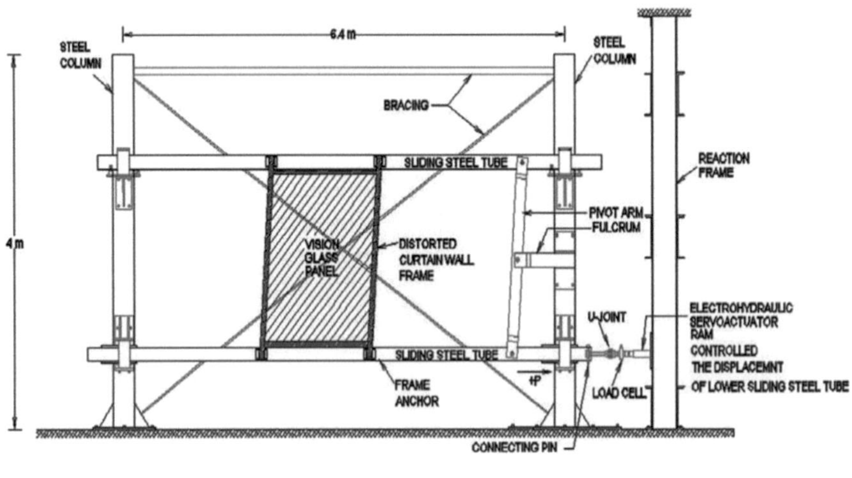

Figure 4.
Schematic drawing of racking test facility designed for mock-up of glass CWs [15].

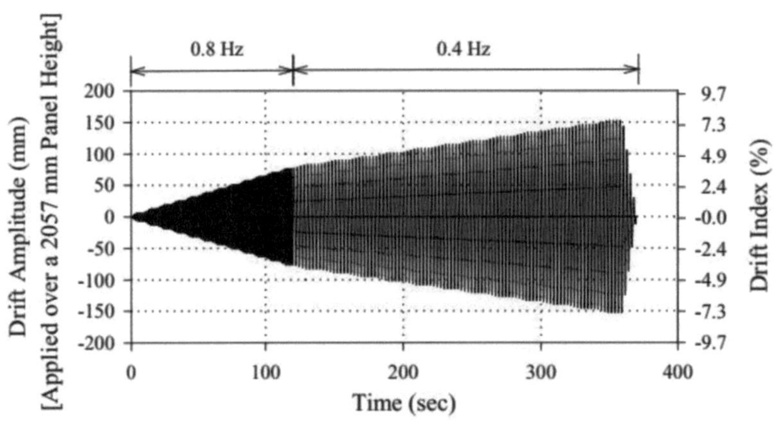

Figure 5.
Standard time-dependent drift pattern for AAMA 501.6 dynamic racking crescendo test.

152.4 mm is reached. These criteria determine when the test concludes. In the case of condition (1), the displacement at which the glass fallout transpires is designated as $\Delta_{fallout}$.

By obtaining this dynamic data and incorporating it within the equations supplied by ASCE 7-16, as explained earlier, one can establish the system capacity to withstand seismic forces while adhering to specific relative drift limits.

4.2 Wind load

Eurocode 1 [31] outlines wind load calculations for façades in part 1–4, aiding structural design for buildings up to 200 m tall and to bridges having no span greater than 200 m. Wind forces are specified for the entire structure and components like cladding units. As a convention, the fundamental basic wind velocity is the 10-minute mean wind velocity with a return period of 50 years. It should be noted that traditional buildings have experienced extreme weather events, revealing the inadequacy of the current guidance, where the design values often fall below actual wind loads observed over decades, as reported in [5] for many regions of Europe, the US, as well as the Asia-Pacific region. On the other hand, many instances still exist where the code fails to offer satisfactory answers. For such situations, wind tunnel experiments or, in exceptional cases, full-scale experiments can provide solutions [32]. Wind tunnel experiments serve as an alternative to codes of practice when dealing with scenarios beyond their scope or when a more precise assessment of wind loading is deemed essential. Within a wind tunnel, the wind, the structure, its environment and occasionally its actions are replicated at a reduced scale. This allows for the measurement of wind speeds, pressures, forces, moments, accelerations and so on. Common test protocols are used to evaluate the performance of glass CWs against out-of-plane loads and weather conditions, both in laboratory and field settings. These protocols, such as ASTM E283 [33], E330/330 M [34], E331 [35], E783 [36], E997 [37] and E1996 [38], cover various aspects including air leakage, structural performance, water penetration, glass breakage probability and impact resistance. These tests are conducted using an air pressurised test chamber as can be seen in **Figure 6** for both glass CW specimens and structural glass panels. Controlled air pressure differentials simulate wind pressures, both static and cyclic, along with additional conditions like debris impacts and water pressures. These conditions aim to replicate realistic scenarios during windstorms.

It is worth mentioning that the design of glass CWs is an open-ended process, requiring engineers to consult various design guides for handling out-of-plane

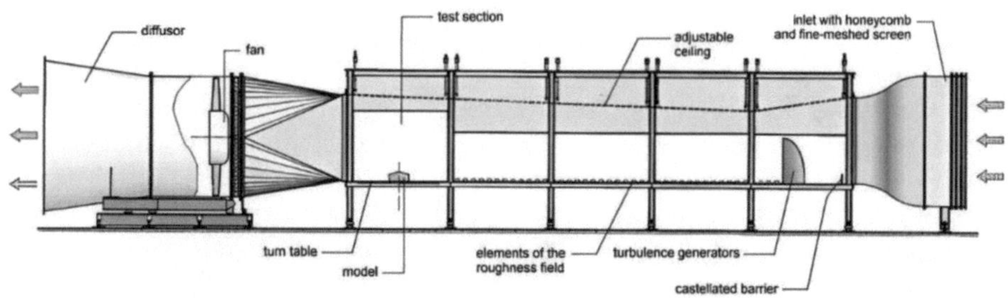

Figure 6.
Schematic of wind tunnel test of the Ruhr University in Bochum, Germany [39].

loading. In addition, the conjunction of wind load with other climate changes like rain [40], hail [41], flood [42] and so on often negatively affects building façades' performance and durability. This includes surface material degradation, frost damage, salt efflorescence, structural cracking, interior harm, and other concerns, and hence, careful consideration should be taken into account for these aspects.

4.3 Impulsive loads

Impulsive loads are sudden and high-intensity forces or pressures applied to a structure within a short duration. These loads can result from events like explosions, impacts or other abrupt occurrences. Impulsive loads are characterised by their rapid rise in force and are typically short-lived but can exert significant stress on the structure [43–45]. Examples of impulsive loads include the shockwaves from explosions, the impact of heavy objects on a surface or the force exerted during a sudden collision. Due to the abrupt nature and intensity of an impulsive load, it can lead to structural damage, material deformation and even failure if not adequately considered in design and engineering. Engineers often need to account for impulsive loads when designing structures and their envelopes that are prone to encountering these sudden and high-energy events [46]. Mitigating the effects of impulsive loads requires careful analysis, materials selection and structural design that can absorb or distribute the impact forces effectively. It is important to note that a majority of casualties resulting from a blast incident are linked to injuries caused by glass fragments [5]. Therefore, extra caution should be exercised when considering glass façades in blast-prone scenarios.

4.3.1 Blast load

The progress in blast protection design achieved during the Second World War resulted in the release of an engineering manual by the United States Army Corps Of Engineers, which is labelled as UFC 3–340-02 [47] after many revisions, and it is widely used by researchers in the field of blast-resistant structures. The manual outlines blast parameter calculation and design techniques for protective construction in facilities involving explosive materials. Its strategies can be adapted for various types of structures directly, or by modifying via experimentation and numerical analyses. Notably, Section 6-27 of this manual focuses on designing glass panes under explosive conditions, providing instructions and graphs based on panel dimensions (thickness, width and height), time duration and blast pressure. These graphical representations are derived to aid in designing and assessing glazing ability to withstand prescribed blast loads with a failure probability not exceeding 0.001. Failure is assumed when maximum deflection of pane exceeds ten times the glazing thickness, preventing edge disengagement of the plate while staying within Von Kármán plate equation limits. In addition, further explanations are given in this manual as design criteria for the glass façades specifically for sealants, gaskets, beads, glazing setting, frame loads and rebound (which is the response to the dynamic loading will cause the window to rebound (outward deflection) after its initial positive (inward) deflection). The most important criteria as maximum allowable limits for frame design are: i) Frame members' relative displacement should be limited to the smaller of 1/264th of the span or 1/8 inch; ii) Maximum stress in any member and fastener should not exceed material yield stress divided by 1.65 and 2.00, respectively, and iii) The deflection of the building should not impose deflections on the frame greater than 1/264th of the length of the pane edge. Also, other codes such as HOSDB, 1997 [48]; GSA TS-01 [49]; ASTM

F1642 [50] and ISO 16933 [51] can be found in the literature regarding glazing systems subject to air blast loadings.

It is important to emphasise that the conventional regulations and building guidelines do not adequately address potential threats that could arise, such as explosive incidents. In order to ensure the protection of constructed infrastructure, there is a need to develop methodologies that can effectively quantify the capacity of structural elements to withstand explosive loads. Additionally, assessing the risks associated with the failure of these elements is crucial. To achieve these goals, a combination of experimental studies and numerical approaches is essential. This combined effort will not only provide practical solutions but also equip engineers with decision-making tools to enhance the security of vital infrastructure. In the realm of testing the blast resistance of glazing materials, two primary methods are commonly employed: shock tube [52] and arena test [51]. Shock tubes are capable of generating relatively moderate pressures over extended durations, making them well-suited for assessing the effects of larger-scale explosive devices, such as vehicle-borne improvised devices and industrial explosions. Conversely, arena tests simulate scenarios involving smaller charges detonated at close range or vehicle-borne improvised explosive devices. Examples of shock tube testing and arena testing of full glass windows under blast loading can be seen in **Figure 7**. It is noteworthy that variations in design codes need to be considered when designing building façades to withstand blast events. In this regard, a comprehensive analysis was conducted to compare the different existing standards for testing blast-resistant windows and glazing materials, as referenced as [55].

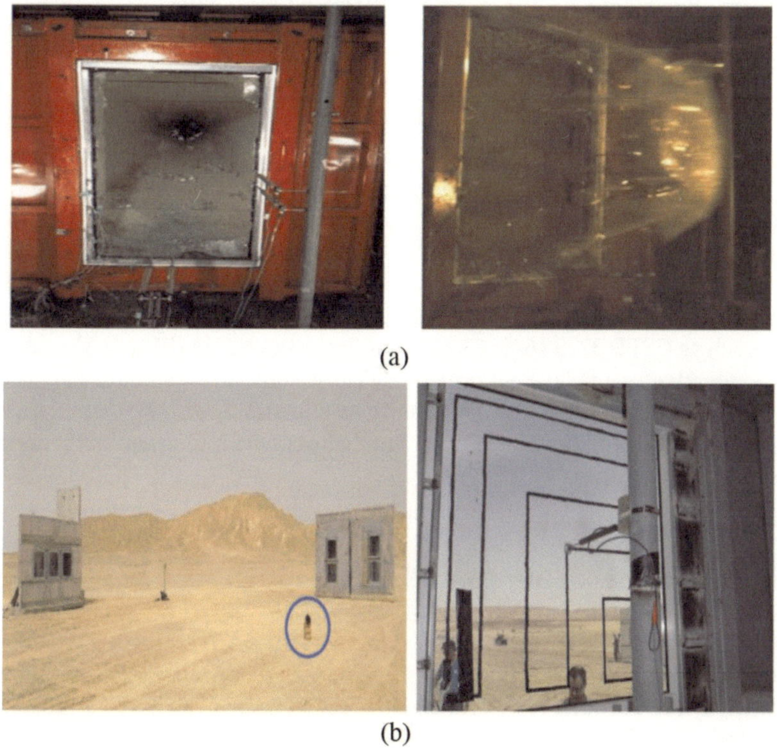

Figure 7.
Examples of testing full glass windows under blast loading; a) shock tube blast testing [53] and b) arena testing [54].

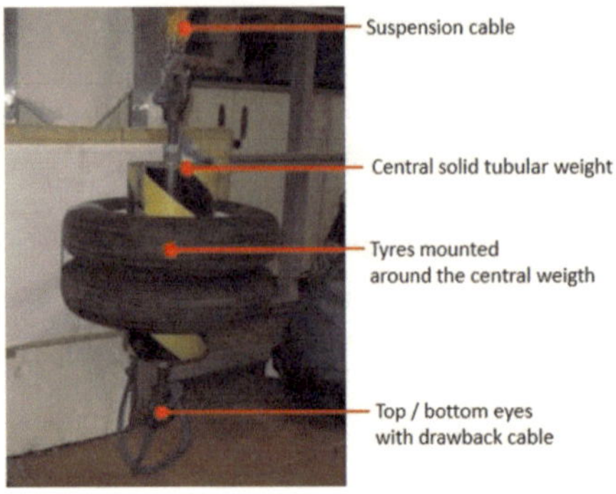

(a)

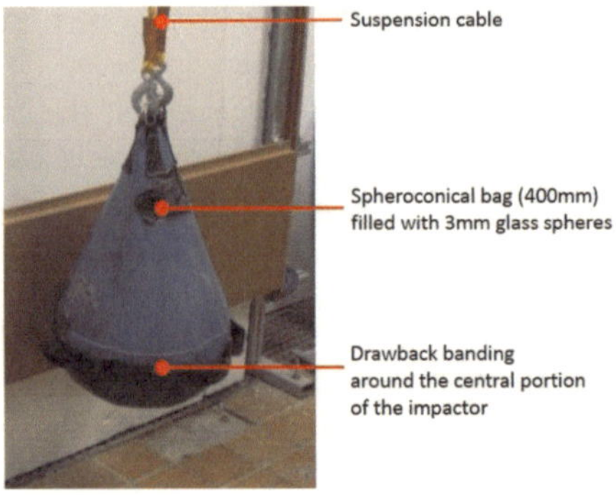

(b)

Figure 8.
Conventional impactors for glass CWs: (a) twin-Tire and (b) Spheroconical bag (SB, with dimensions in mm) [61].

4.3.2 Impact load

Impact loads represent a prevalent form of dynamic forces that can inflict significant damage on brittle materials in a short duration of time. Present regulations do not have enough guidance to accurately assess the resistance of glass elements to various types of impact loading. The classification of impact loading, as outlined in [56], distinguishes between hard body and soft body impact, a division particularly pertinent to glass due to its vulnerability when interacting with harder materials. However, there are only a limited number of regulations such as HR EN 12600 [57], HR EN 356 [58], DIN 18008-4 Annex A document [59] and CWCT TN 76 [60] that detail methods for testing glass resistance to impact. Two commonly used impactors are the spheroconical bag (SB) and the twin tire (TT) which are widely used to assess the performance of glazing under soft body impact. The SB contains glass spheres and weighs 50 kg, while the TT consists of two pneumatic tires inflated to 3.5 bar air pressure, with an additional 50 kg steel mass inside (see **Figure 8**). The International HR EN 12600 standard introduced the TT pendulum protocol to replace the SB impactor. German regulations also permit FE numerical simulations using the TT impact instead of full-scale experiments. Some standards such as CSTB 3228 [62], CWCT TN 76 [60], ACR[M]001 [63] and ANSI Z97.1 [64] still advocate for the SB impactor. These standards evaluate the glass system performance after impact, assessing its ability to withstand breakage, cracks and fragments. The primary regulations influence façade designers and manufacturers to adhere to the original SB approach.

5. Numerical analysis of in-plane seismic load effects

As for many other constructional and structural issues, engineers have consistently turned to numerical methods, encompassing simpler techniques or complicated finite element analysis, to account for intricate nuances in their models. In addition to mitigating the financial burden associated with laboratory testing, these numerical methods facilitate parametric studies involving diverse input variables.

A huge number of research studies can be found in the literature regarding performance evaluation of glass elements and façades under different types of loadings.

There are different methods including single degree of freedom (SDOF), multi degree of freedom (MDOF) and finite element (FE) methods that are widely used to find the response of such structures. SDOF methods are always used to find the response of a single member under extreme dynamic loads like blast and impacts [65, 66], while MDOF [67, 68] and FE methods [69–73] not only can be used for single members but also can be utilised for other complicated structures where more details should be taken into consideration. Besides, engineers are always seeking simpler solutions (instead of experimental and numerical analyses) to solve problems that can provide more straightforward analytical approaches for investigating the issue. As a result, analytical methods (which themselves have been validated and calibrated using accurate experimental and numerical results) have also gained importance among researchers, and examples of these can be observed in [16, 74, 75]. Exploring this topic in depth within the confines of this chapter is not feasible; however, more comprehensive explanations can be sought in the technical literature.

In the following, to show the accuracy and applicability of FE models to find the response of a glazing under lateral loading, a numerical model based on Abaqus software is used to evaluate the response of a dry-glazed CW façade under lateral load, based on validation towards literature tests. In this regard, the experimental investigation conducted by Shirazi [76] is taken into account. The schematic of selected experimental test is shown in **Figure 9**. The CW arrangement depicted in **Figure 9**

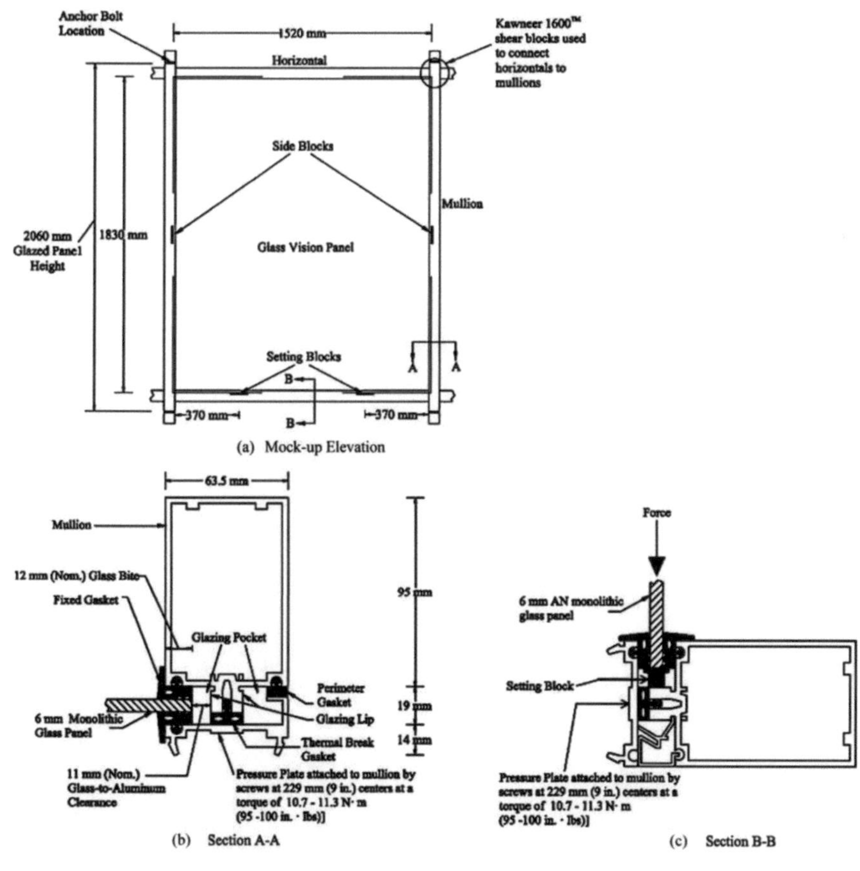

Figure 9.
Considered glass CW configuration for mock-up test [71].

features an annealed monolithic glass panel measuring 1829 mm in height, 1524 mm in width, and 6 mm in thickness, which was dry glazed within a Kawneer 1600 aluminium CW frame. It is important to mention that the clearance between the glass and the frame was taken as 11 mm. The pressure plate profiles are affixed to the mullions and transoms using screws spaced at intervals of 9 inches centre-to-centre, and these screws are tightened to a required torque of 10.7–11.3 N-m to secure the glass in position. The CW was subjected to a lateral racking displacement at the top corner of the frame.

To simulate the glass CW, all components (including mullions, transoms, pressure plates, gaskets, perimeter gaskets, thermal gaskets, setting blocks and glass panel), are individually modelled and meshed using C3D8R elements. The particular aspect of the present application – compared to a multitude of literature examples which are based on the use of rough geometrical simplifications – is in fact a full three-dimensional description of CW elements. These parts are then assembled to construct the final configuration. It is important to highlight that the minimum dimension of solid elements is 2 mm, which is primarily applied to gaskets, where higher deflection is anticipated. **Figure 10** shows the CW modelled in Abaqus software.

The FE modelling takes into account various interactions among distinct components, encompassing the interactions between i) glass, transoms and mullions; ii) glass and gaskets; iii) glass and setting blocks; iv) rubber parts (i.e. gaskets, perimeter gaskets, thermal gaskets, setting blocks) and aluminium parts (i.e. mullions, transoms, pressure plates) and v) the semi-rigid connections linking transoms and mullions. In cases (i) and (iii), the normal hard contact and frictionless tangential behaviours are used to define the contact property. In case (ii), the hard contact is used for normal behaviour, while the penalty method with a friction coefficient of 0.65 is used to define the tangential behaviour of contact property. In case (iv), the tie constraint strategy is used, and for case (v), the u-joint connection type is implemented to define

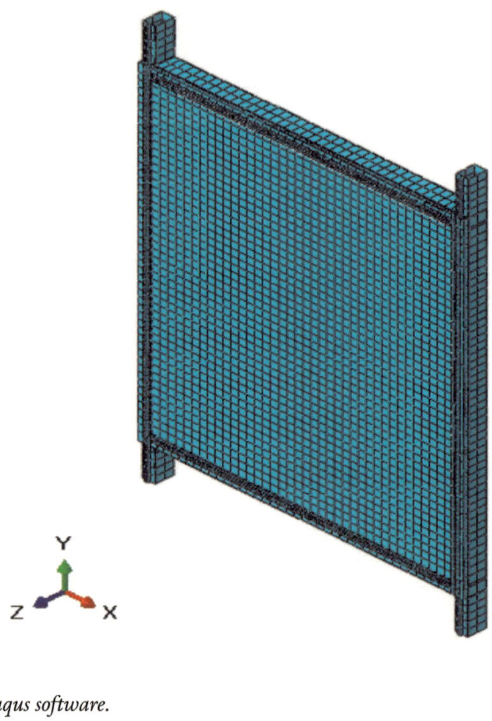

Figure 10.
Glass CW modelled in Abaqus software.

the connector in connecting transom and mullion parts. The U-joint type connector, which is used to connect two reference points at each location (i.e. the connection of the transom to the mullion) by the wire, fixes all translational degrees of freedom and the rotational degree of freedom about the global y-axis.

The glass panel characteristics are defined with a modulus of elasticity of 72 GPa, Poisson ratio of 0.25 and mass density of 2500 kg/m^3. For the aluminium parts, values of 69 GPa for the modulus of elasticity, 0.33 for Poisson ratio, and 2700 kg/m^3 for mass density are utilised. Regarding the gaskets, the modulus of elasticity is approximated at 4.4 MPa, with Poisson ratio equal to 0.3 and mass density of 1300 kg/m^3.

The results of FE analysis are compared with the reference experimental results in terms of load-drift relationship, as demonstrated in **Figure 11**. The figure reveals that there is a rather good agreement between the FE outcomes and the experimental findings. In other words, this signifies that the FE modelling can anticipate the structural responses at particular drift levels during the analysis, in accordance with the experimental findings. These structural behaviours encompass three essential aspects (which are shown in **Figure 11**) including: 1) starting the plastic deformation of the gasket, 2) starting the contact between the glass and the frame and 3) the occurrence of frame and glass failure. Furthermore, the model proficiency in faithfully replicating the glass movement within the glazing pocket enables accurate representation of how the glass comes into contact with the frame, and all these aspects have a kay role in structural performance and capacity assessment for similar systems.

Assured that the FE model like in **Figure 10** can be further optimised and simplified to enhance its computational efficiency, it is worth noting that major challenges are related to the accuracy of simplified restraints and boundaries, given that they have a major influence on the stress and strain distribution in glass. Obviously, special modelling assumptions should be taken into account under various loading configurations. However, the goal of ongoing investigations is to capture and define a harmonised modelling strategy for glass façades in general. Also, another important aspect which is presently under investigation is the possible definition of standardised performance indicators that could be used for a given curtain wall exposed to various

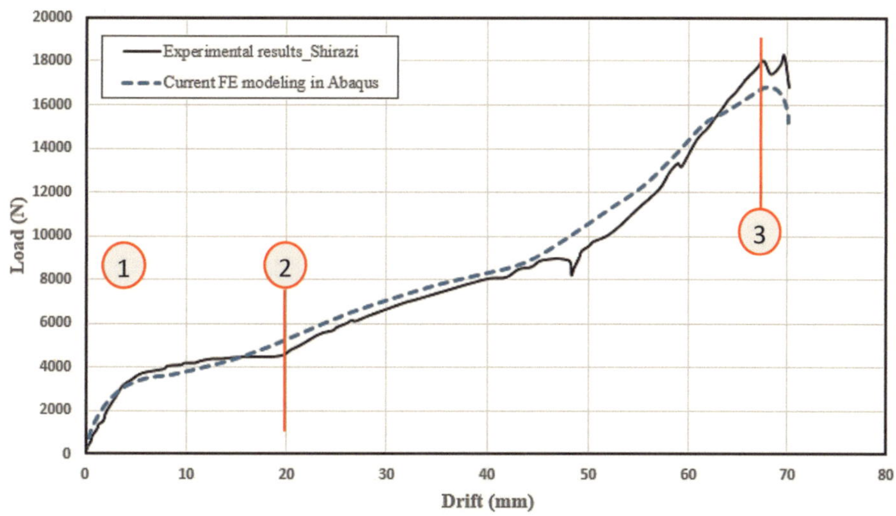

Figure 11.
Comparison of load-drift relationship obtained from FE model with experiment.

mechanical loads and for structural health monitoring purposes, diagnostics and life cycle assessment of façades in buildings.

6. Conclusion

A comprehensive exploration of glass curtain wall systems under various dynamic loads, including seismic, wind and impulsive forces like blasts and impacts, is provided by this paper. Regulations, experimental methods and numerical simulations have received special attention, and the essential need for meticulous design, advanced materials and rigorous testing to ensure the structural integrity and safety of these architectural features is emphasised. The array of design standards, regulations and codes available to guide engineers in addressing these dynamic loads is illuminated by the paper. The necessity of a multidisciplinary approach, encompassing elements of structural engineering, material science and architectural design to create glass curtain walls capable of withstanding dynamic loads, is highlighted. Additionally, the role of numerical methods, particularly finite element analysis, in simulating and predicting the behaviour of glass curtain wall systems under dynamic conditions, is underscored by the study. These methods offer cost-effective and efficient means of assessing complex interactions, enabling the evaluation of structural responses and contributing to design optimization. As the boundaries of modern architecture continue to be pushed by architects and engineers, the insights presented here serve as a valuable guidance to ensure the optimal performance of glass curtain wall structures when confronted with dynamic challenges.

Author details

Mohammad Momeni and Chiara Bedon*
Department of Engineering and Architecture, University of Trieste, Italy

*Address all correspondence to: chiara.bedon@dia.units.it

References

[1] Ryan P, Otlet M, Ogden R. Steel Supported Glazing Systems. Steel Construction Inst; 1997

[2] Gil EM. Computational Modeling of Glass Curtain Wall Systems to Support Fragility Curve Development. Virginia Tech; 2019

[3] Memari AM, Schwartz TA. Glazing and curtain wall systems to resist earthquakes. Architectural Glass to Resist Seismic and Extreme Climatic Events. 2009:28-63

[4] Elkadi H. Cultures of Glass Architecture. Routledge; 2016

[5] Bedon C et al. Performance of structural glass facades under extreme loads–design methods, existing research, current issues and trends. Construction and Building Materials. 2018;**163**:921-937

[6] Louter C, Bedon C, Kozłowski M, Nussbaumer A. Structural response of fire-exposed laminated glass beams under sustained loads; exploratory experiments and FE-simulations. Fire Safety Journal. 2021;**123**:103353

[7] ASTM. Standard Specification for Heat-Strengthened and Fully Tempered Flat Glass. ASTM C 1048; 2012

[8] Haldimann M, Luible A, Overend M. Structural Use of Glass. Iabse; 2008

[9] Zhang X, Hao H. The response of glass window systems to blast loadings: An overview. International Journal of Protective Structures. 2016;**7**(1):123-154

[10] Zhang X, Zou Y, Hao H, Li X, Ma G, Liu K. Laboratory test on dynamic material properties of annealed float glass. International Journal of Protective Structures. 2012;**3**(4):407-430

[11] E. Team. Northridge earthquake reconnaissance report. Earthquake Spectra. 1996:49-98

[12] Neyraval P, Egan SM, Basco J, Arques JS. Explosion of 14th January 2020 at IQOXE, Tarragona, Spain. Chemical Engineering Transactions. 2022;**90**:649-654

[13] Sivaraman S, Varadharajan S. Investigative consequence analysis: A case study research of Beirut explosion accident. Journal of Loss Prevention in the Process Industries. 2021;**69**:104387

[14] A. Minimum Design Loads for Buildings and Other Structures. American Society of Civil Engineers (ASCE); 2016. pp. 7-16

[15] Memari AM. Curtain Wall Systems: A Primer. American Society of Civil Engineers (ASCE); 2013

[16] Caterino N. A simplified analytical approach for assessing non-linear in plane seismic response of stick-built glass curtain walls. In: Structures. Vol. 49. Elsevier; 2023. pp. 281-294

[17] UNI EN 1994-1-1:2005 – Eurocode 4 – Design of Composite Steel and Concrete Structures

[18] Zealand SN. Structural Design Actions, Part 5: Earthquake Actions. NZS1170. 5; 2004

[19] Seminar C. Nehrp Guidelines for the Seismic Rehabilitation of Buildings (Fema 273). Washington, DC, USA: Building Seismic Safety Council; 1997

[20] A. T. Council and N. E. H. R. Program. Seismic Performance Assessment of Buildings. Federal Emergency Management Agency; 2012

[21] Cremen G, Baker JW. Improving FEMA P-58 non-structural component fragility functions and loss predictions. Bulletin of Earthquake Engineering. 2019;**17**:1941-1960

[22] Del Gobbo GM, Williams MS, Blakeborough A. Seismic performance assessment of Eurocode 8-compliant concentric braced frame buildings using FEMA P-58. Engineering Structures. 2018;**155**:192-208

[23] S. U. Construction and C. Commission. Technical code for building curtain wall. In: DGJ08-56-2012/J12028–2012. Shanghai, China: Shanghai Construction; 2012

[24] J. 102. Technical Code for Glass Curtain Wall Engineering. China: China Architecture & Building Press Beijing; 2003

[25] C. S. Association. Guideline for Seismic Risk Reduction of Operational and Functional Components (OFCs) of Buildings. Canadian Standards Association; 2001

[26] A. JASS14. Japanese architectural standard specification curtain wall. AIoJ (AIJ). 1996

[27] I. C. Council. Acceptance Criteria for Seismic Qualification by Shake-Table Testing of Nonstructural Components and Systems. Whittier CA: International Code Council AC156; 2007

[28] A. T. Council, M.-A. E. Center, M. C. f. E. E. Research, P. E. E. R. Center, and N. E. H. R. Program. Interim Testing Protocols for Determining the Seismic Performance Characteristics of Structural and Nonstructural Components. Federal Emergency Management Agency; 2007

[29] A. A. M. Association. Recommended Static Test Method for Evaluating Curtain Wall and Storefront Systems Subjected to Seismic and Wind Induced Interstory Drifts. Vol. 1. Publication No. AAMA 501.4; 2001

[30] A. A. M. Association. Recommended Dynamic Test Method for Determining the Seismic Drift Causing Glass Fallout from a Wall System (AAMA 501.6–09). Schaumburg, United States; 2009

[31] E. 1991-1. Eurocode 1 – Actions on structures. Brussels, Belgium: European Committee for Standardisation (CEN); 2005

[32] Stathopoulos T, Baniotopoulos CC. Wind Effects on Buildings and Design of Wind-Sensitive Structures. Springer Science & Business Media; 2007

[33] E. ASTM. 283-rate of air leakage through exterior windows. Curtain Walls and Doors Under Specified Pressure Difference Across the Specimen

[34] ASTM, editor. Standard Test Method for Structural Performance of Exterior Windows, Doors, Skylights and Curtain Walls by Uniform Static Air Pressure Difference. West Conshohocken, PA: ASTM; 2009

[35] E. ASTM. 1105–00. Standard Test Method for Field Determination of Water Penetration of Installed Exterior Windows, Skylights, Doors, and Curtain Walls by Uniform or Cyclic Static Air Pressure Difference. West Conshohocken, PA: ASTM International

[36] ASTM. Standard test method for field measurement of air leakage through installed exterior windows and doors. 2010

[37] E. ASTM. Test method for evaluating glass breakage probability under the influence of uniform static loads by proof load testing. 2015

[38] ASTM E. Specification for performance of exterior windows. Curtain Walls, Doors and Impact Protective Systems Impacted by Windborne Debris in Hurricanes. 1996

[39] Geurts C, van Bentum C. Wind loading on buildings: Eurocode and experimental approach. In: Wind Effects on Buildings and Design of Wind-Sensitive Structures. Springer; 2007. pp. 31-65

[40] Gholamalipour P, Ge H, Stathopoulos T. Wind-driven rain (WDR) loading on building facades: A state-of-the-art review. Building and Environment. 2022;**221**:109314

[41] Paterson DA, Sankaran R. Hail impact on building envelopes. Journal of Wind Engineering and Industrial Aerodynamics. 1994;**53**(1-2):229-246

[42] Haghighatgou N, Daniel S, Badard T. A method for automatic identification of openings in buildings facades based on mobile LiDAR point clouds for assessing impacts of floodings. International Journal of Applied Earth Observation and Geoinformation. 2022;**108**:102757

[43] Kumar A, Saha SK, Matsagar VA. Stochastic response analysis of elastic and inelastic systems with uncertain parameters under random impulse loading. Journal of Sound and Vibration. 2019;**461**:114899

[44] Momeni M, Hadianfard MA, Bedon C, Baghlani A. Damage evaluation of H-section steel columns under impulsive blast loads via gene expression programming. Engineering Structures. 2020;**219**:110909

[45] Momeni M, Bedon C, Hadianfard MA, Baghlani A. An efficient reliability-based approach for evaluating safe scaled distance of steel columns under dynamic blast loads. Buildings. 2021; **11**(12):606

[46] Bedon C. Research developments on glass structures under extreme loads. Modern Trends in Research on Steel, Aluminium and Composite Structures. 2021;**1**:3-13

[47] DoD U. Structures to Resist the Effects of Accidental Explosions. US DoD, Washington, DC, USA. UFC 3-340-022008

[48] Guide GH. Security Facilities Executive Special Services Group—Explosion Protection. Cabinet Office; 1997

[49] A. GSA. Standard Test Method for Glazing and Window Systems Subject to Dynamic Overpressure Loadings. Washington, DC: General Service Administration; 2003

[50] ASTM. Standard test method for glazing and glazing systems subject to airblast loadings. 2012

[51] ISO. Glass in building—Explosion-resistant security glazing—Test and classification for arena air-blast loading. 2007

[52] I. 16934. Glass in building—Explosion resistant security glazing—Test and classification by shock tube loading. 2007

[53] Blast Resistant Window Anchor Systems. Available from: http://www.ericjacques.com/project/blastresistantanchors/

[54] Seica MV, Packer JA, Yankelevsky DZ. Blast and impact loading effects on glass and steel

elements and materials. Thin-Walled Structures. 2019;**134**:384-394

[55] Bedon C et al. A comparison of existing standards for testing blast resistant glazing and windows. European Commission, Joint Research Centre, Institute for the Protection and Security of the Citizen. 2014

[56] Feldmann M, Kasper R, Abeln B, Cruz P, Belis J, Beyer J. Guidance for European structural design of glass components. Publications Office of the European Union. 2014:1-196

[57] Grozdanić G, Galić M, Marović P. Some aspects of the analyses of glass structures exposed to impact load. Coupled Systems Mechanics. 2021;**10**: 475-490

[58] Institution BS. Glass in Building-Security Glazing-Testing and Classification of Resistance against Manual Attack. British Standards Institution; 2000

[59] Siebert G. DIN 18008 Glas im Bauwesen–Bemessungs-und Konstruktionsregeln: Grundlagen und Konzept der Normenreihe. Stahlbau. 2019;**88**(1):32-35

[60] C. T. 76. Technical Note 76: Impact Performance of Building Envelopes: Method for Impact Testing Cladding Panels. Bath, UK: Centre for Window & Cladding Technology (CWCT); 2012

[61] Bez A, Bedon C, Manara G, Amadio C, Lori G. Calibrated numerical approach for the dynamic analysis of glass curtain walls under spheroconical bag impact. Buildings. 2021;**11**(4):154

[62] C. C. 3228. Resistance Impact on Glass and Glass Roof Infill. Marne-laVallée, France: Centre Scientifique et Technique du Bâtiment (CSTB); 2014

[63] ACR[M]001. Test for Non-Fragility of Profiled Sheeted Roof Assemblies. London, UK: Advisory Committee for Roofwork (ACR); 2014

[64] A. Z97.1. American National Standard for Safety Glazing Materials Used in Buildings—Safety Performance Specifications and Methods of Test. McLean, VA, USA: Accredited Standards Committee (ASC); 2015

[65] Feng R-Q, Ye J-H, Wu Y, Shen S-Z. Nonlinear response spectra of cable net facades. Soil Dynamics and Earthquake Engineering. 2012;**32**(1):71-86

[66] Ding C, Ngo T, Mendis P, Lumantarna R, Zobec M. Dynamic response of double skin façades under blast loads. Engineering Structures. 2016;**123**:155-165

[67] dos Santos FA, Cismasiu C, Bedon C. Smart glazed cable facade subjected to a blast loading. Proceedings of the Institution of Civil Engineers-Structures and Buildings. 2016;**169**(3):223-232

[68] Gupta S, Banfill B, Matsagar V. MDOF Modeling and blast dynamic behavior of Curtain Wall with variable damping approach. Journal of Structural Engineering. 2023;**149**(9):04023122

[69] Bedon C, Amadio C. Numerical assessment of vibration control systems for multi-hazard design and mitigation of glass curtain walls. Journal of Building Engineering. 2018;**15**:1-13

[70] Bedon C, Amadio C. Enhancement of the seismic performance of multi-storey buildings by means of dissipative glazing curtain walls. Engineering Structures. 2017;**152**:320-334

[71] Memari AM, Shirazi A, Kremer PA, Behr RA. Development of finite-element modeling approach for lateral load

analysis of dry-glazed curtain walls. Journal of Architectural Engineering. 2011;**17**(1):24-33

[72] Caterino N, Del Zoppo M, Maddaloni G, Bonati A, Cavanna G, Occhiuzzi A. Seismic assessment and finite element modelling of glazed curtain walls. Structural Engineering and Mechanics. 2017;**61**(1):77-90

[73] Aiello C, Caterino N, Maddaloni G, Bonati A, Franco A, Occhiuzzi A. Experimental and numerical investigation of cyclic response of a glass curtain wall for seismic performance assessment. Construction and Building Materials. 2018;**187**:596-609

[74] Bedon C, Amadio C, Sinico A. Numerical and analytical investigation on the dynamic buckling behavior of glass columns under blast. Engineering Structures. 2014;**79**:322-340

[75] Bedon C, Amadio C. Analytical and numerical assessment of the strengthening effect of structural sealant joints for the prediction of the LTB critical moment in laterally restrained glass beams. Materials and Structures. 2016;**49**:2471-2492

[76] Shirazi A. Development of a Seismic Vulnerability Evaluation Procedure for Architectural Glass Curtain Walls. Ph.D. thesis. University Park, PA: Pennsylvania State Univ.; 2005

Chapter 3

The State of Housing, Drinking Water, Electricity, and Sanitation Facilities of Scheduled Tribes in Eastern Uttar Pradesh, India

Poonam Singh Kharwar, Devesh Kumar, Abhishek Kumar and Abhinav Kumar

Abstract

Façade design, drinking water, electricity, and sanitation are critical basic human needs for a decent life in the modern period. The development and implementation of these regulations are necessary for socioeconomic advancement and protect tribes, particularly women, from significant public health, environmental, and security issues. Despite the government's intentions to address their backward status through special constitutional provisions, tribes in eastern Uttar Pradesh remain severely underserved regarding these services. The design of façades has a favorable impact on the lives of socioeconomically deprived citizens of developing countries like India. The present chapter examines the façade design, drinking water, electricity, and sanitation services provided to Scheduled Tribes in the eastern Upper Peninsula and potential improvement initiatives. Façade design impacts the types of businesses that thrive in a given location. The majority of scheduled tribes rely on the informal economy for a living. The majority of ST families (43.9%) still live in jhuggis, only 27.12% have both tap water supplies and electricity, the majority (92.15%) use hand pumps for drinking water outside the home, 77.4% of STs do not have latrine facilities inside the premises, and the surrounding sanitation is inadequate. Although government is taking steps for piped water supply, ST families are still deprived of this facility due to the scattered nature of remotely placed *kaccha* houses and lack of proper attention from responsible authorities.

Keywords: scheduled tribes, housing, sanitation, drinking water, electricity facility

1. Introduction

Tribes are different from the general population because of their different way of living and community life. Tribes as custom-bound communities in India are facing numerous problems like geographic separation from mainstream of the masses such as unemployment, poverty, poor health, alcoholism, various kinds of exploitations, natural calamities, and naxalism in present global area. Their inclusive growth can be achieved only by bringing them into the national mainstream and, at the same time, preserving their culture and traditions. Nomenclature of the Scheduled Tribe (ST) fully emerged under the Government of India Act of 1935 and the Constitution of India to bring them into the mainstream of national development by their equitable and balanced progress [1].

Except for Africa, India has the highest concentration of tribal people worldwide. According to the 2011 census, tribes made up approximately 8.6% of the total Indian population, with 89.97% of them living in rural areas. Uttar Pradesh (U.P.) is the most populous state. However, it is also one of the least developed, with a Human Development Index (HDI) value of 0.380 (2007–2008), lower than the national average of 0.467, placing it 18th among Indian states/U.T.s. It ranks 17th among all Indian states regarding the number of S.T.s. According to the 2011 census, 1.13 million indigenous people made up 0.6% of the overall population of the U.P. The eastern section of the state was home to around 84% of the population. Sonbhadra district accounted for more than one-third (33.94%) of the total S.T. population, while Ballia and Deoria accounted for more than half (53.34%) (2013) (Government of India) [2]. Although HDI improvement in India was estimated to be greater (0.633) in 2021, it was relatively slow (0.592) in Uttar Pradesh, which ranked 32nd out of 33 states [3].

At present, there are 15 notified tribal communities in U.P. *Gond* tribe accounts for first highest number of tribes followed by Kharwar as second most populous tribal community. After 2002 proclamation, Gonds were categorized as ST in 13 districts only, and in other districts, they were renamed SC. In 2011, the Gonds, along with the sub-ethnic groups namely Dhuria, Nayak, Ojha, Pathari, and Rajgond, were the largest and most prominent tribal population, accounting for 50.2% of all STs and occupying 18 districts in eastern Uttar Pradesh. According to the 2011 census, Kharwar is the second most populous tribe, accounting for 14.6% of the state's ST population. These two tribes accounted for nearly two-thirds of the total ST population in the Upper Peninsula. Tharu is the third largest community, with a population increase of 26% from 83,544 in 2001 to 1,05,291 in 2011. Their percentage share of all STs has declined from 77.4 in 2001 to 9.3 in 2011. *Saharya* is the fourth largest tribe, found mainly in Lalitpur district accounting for 6.25 of all STs followed by Chero (3.7%). Thus, according to census 2011, all these five tribes constitute 83.6% of ST population of U.P. *Baiga and Pankha/Panika* constituted 1.5 and 1.4%, respectively. *Agariya and Bhuiya/Bhuinya* constituted 2.6 and 2.2%, respectively. Population share of Bhotia (0.5%), Buksa (0.4%), Janusari (0.3%), Raji (0.1%), Parahiya (0.1%), and Patari (0.01%) contributes to minimum in ST population. Sonbhadra district constituted more than one-third (33.94%) and along with Ballia and Deoria more than one-half (53.34%) of total ST population of the U.P. [4].

Welfare programs directed for the development of these STs have not resulted in any visible positive impact. Given the common backwardness and suffering of the S.T. people in the eastern U.P., it is critical to study and uncover the underlying correlates that make their lives so wretched. Façade design, drinking water, electricity, and sanitation are crucial basic human necessities for a decent life in the modern day, and the development and implementation of these provisions are critical for the socioeconomic uplift of these less fortunate parts of society. Population of STs is less in eastern U.P. So it has not drawn attention of researchers in the past. No extensive field study has been reported on STs in socioeconomically backward regions of eastern U.P. Hence, there is an urgent need to conduct such study to fill up the gap of knowledge and to provide guidelines and strategies for formulation of sustainable development program for the overall betterment of these deprived community.

2. Objectives

Present study was conducted among families of different ST communities in Deoria, Ballia, Ghazipur, Varanasi, and Sonbhadra districts of U.P. with following objectives:

1. To study the status of housing in terms of types of houses and locality used for living by STs in eastern U.P.

2. To study the facilities of tap water supply and electricity for STs in eastern U.P.

3. To study the drinking water supply source, distance, and purification used by STs in eastern U.P.

4. To study the availability of sanitary latrine and sanitation status in ST families of eastern U.P.

5. To compare the housing, drinking water, electricity, and sanitation status of STs with general population of eastern U.P.

6. To explore the further possibilities to strengthen housing, drinking water, electricity, and sanitation facilities in ST communities in eastern U.P.

3. Review of literature

Belshaw [5] mentioned that though a lot has been done for tribal' s social and economic betterment, a great deal remains to be done.

Sharma [6] mentioned that 20% of the tribal population has been uprooted and displaced in less than 50 years; they have lost their rights because of their political powerlessness. The magnitude of land and number of displaced persons has been increasing since then.

Sharma [6] mentioned that 20% of the tribal population has been uprooted and displaced in less than 50 years; they have lost their rights because of their political powerlessness. The magnitude of land and number of displaced persons has been increasing since then.

According to Singh [1], the tribals in India are the most adversely impacted ethnic group due to post-independence development, and the new economic policy is likely to worsen their situation. As a result, more earnest efforts are required to salvage and enhance their socioeconomic situation within the restrictions and possibilities of their existential circumstances, including rural, illiteracy, poverty, ill-health, and unproductive agriculture. The government's efforts to improve tribal welfare through protective developmental measures have had little impact on tribal development. Mehta [7] gave a comprehensive analysis of tribal development initiatives used during the twentieth century, revealing that the government failed to provide them with basic survival needs.

Mondal and Mete [8] noted that tribes are not able to appreciate modern concept of health and sanitation due to illiteracy and ignorance. Based on NITI Aayog estimates (2011–2012), U.P. stands among states having 30–40% population below poverty line and it is better only to Jharkhand and Chhattisgarh. Deshmukh [9] revealed that the existing welfare strategies did not help the tribal overcome from inferiority and atrocities on them.

Although welfare plans such as subsidizing housing (façade designed by the government as multistory building) like *Lohia, Indira, and Kashiram Awas Yojna* exist for the poor in rural area, tribes are not getting benefits and they are victims of inequality, exploitation, and oppression. Their economic situation is worse than other communities in society, the majority of them are deprived of the basic needs of life. Compared to urban areas, situation of tribal living in remote area is worse [10].

Access to adequate drinking water and sanitation services is intimately linked to public health. Consumption of contaminated drinking water, poor disposal of human excreta, a lack of personal and food hygiene, and improper solid and liquid waste disposal have all been identified as causes of numerous diseases in developing nations such as India. According to the 2011 Census of India, almost 70% of India's population (650 million) lives in rural and slum areas. It increases the population's vulnerability to water-borne and vector-borne diseases. It is also due to a need for more basic sanitation facilities, contaminated water, and unsanitary living conditions.

According to the 2011 census, 40.62% of STs live in good-conditioned houses with sustainable façade designs. Meanwhile, 6.2% live in crumbling façade-designed dwellings, compared to 53.1% and 5.35% for all socioeconomic groups. The availability of drinking water portrays a dismal perception, with only 19.72% of STs having a drinking water source inside their premises and 33.59% having it outside their premises. The other group does better (46.6 and 17.6%, respectively). The hand pump is the primary drinking water source for STs and all categories—all categories (33.5%) and STs (39.2%). Tap water from treated sources is the second most available source for all social group households (32%), whereas in case of STs, it is water from uncovered wells (19.1%).

In India, 77.4% of STs do not have access to a latrine, compared to 53.1% of the general population. Only 46.9% of all homes, including 22.6% of ST households, have a latrine. Human night soil removal is still used by up to 0.3% of all households and 0.1% of ST households. While just 49.8% of total households utilize open defecation, 74.7% of ST households still practice it [2].

The disparity in terms of access to household amenities like tap water and latrine is sharp across states. While facility of tap water is as high as 89.5% in Himachal Pradesh and 85.4% in Sikkim and Goa, it is only 27.35% in U.P. Facility of drinking water within the premises is as high as 85.9% in Punjab, it is only 51.9% in U.P. The government initiated a new project supported by the World Bank called as National Rural Drinking Water Programme, which aims to provide safe, 24 × 7 piped drinking water supply to 7.8 million rural population in four low-income states namely Assam, Bihar, U. P., and Jharkhand that have the lowest piped water supply and sanitation facilities.

While access to and coverage of latrine facilities is only 35.7% in Uttar Pradesh, it also attempts to promote excellent hygiene and cleanliness among people by launching Solid and Liquid Waste Management initiatives in villages, towns, and cities. Since the commencement of the Swachh Bharat Mission, sanitation progress has accelerated. According to the NSSO, sanitation coverage has increased to roughly 48.8% as of December 2015. The mission's intended outcomes are the maintenance of installed toilets and their use by beneficiaries [11].

Jaiswal [12] found that more than 55% of tribes stay in *kuccha* houses façade design made of mud and natural local amenities, half of the population lack pure water; more than 60% tribal areas are not electrified.

Bano and Ara [13] found only 22% literacy in Kharwar tribe of Sonbhadra, the majority of them (66%) were agriculturists followed by labor work (30%), mostly (92%) living in semicemented houses. Their economic status was pulling them down due to backwardness in education, lack of ideas and techniques, lack of knowledge and skill production, and inability to manage their income.

4. Methodology

All ST communities living in eastern U.P. comprised as universe of the study. Five districts of eastern U.P. namely Sonbhadra, Varanasi, Ghazipur, Deoria, and Ballia were selected randomly to conduct the study. With the assistance of health specialists and related experts, the author created a semiconstructed questionnaire based on perspectives regarding general health, education, and socioeconomic level markers. Section A contains 22 socioeconomic status questions developed by Aggarwal et al. [14], a scale suitable for all segments of society. Section B includes questions about the general health and education of the head of the household and family members. Section C contains questions about general health, education, and socioeconomic status variables. Present research project, being extensive field study, was performed by survey research method based on the primary as well as secondary data collected by observation and interview. Field surveyors used the semiconstructed questionnaire to collect the data from the study sample, which consists of selected 11,416 families residing in 474 villages of five districts. Field surveyors also subjected them to scheduled information interviews and observation techniques as needed. The secondary data were collected from the relevant published documents. Data were compiled in Excel sheet of SPSS version 16, analyzed, and subjected to vigorous statistical treatment for analysis as needed.

5. Results

5.1 Housing

5.1.1 Type of house used for living in family

Analysis of different types of houses used for living by families in study districts is presented in **Figure 1**. Most of them live in either *jhuggis* (43.9%) or own houses with 1–2 rooms (45.5%), and only 10.5% own houses with 3–4 rooms but none have five or more room houses.

5.1.2 Facilities of tap water supply and electricity

Analysis of tap water supplies and electricity in families in study districts is presented in **Figure 2**. It reveals that only 27.12% of ST families have both tap water supply and electricity and 43.7% have none of it. Sonbhadra families are without tap water supply and electricity connection is limited to 22.5% only.

5.1.3 Drinking water supply source, distance, and purification status

Analysis of drinking water supply and purity status of families in study districts is presented in **Table 1**. Most of the ST families (92.15%) collect drinking water directly

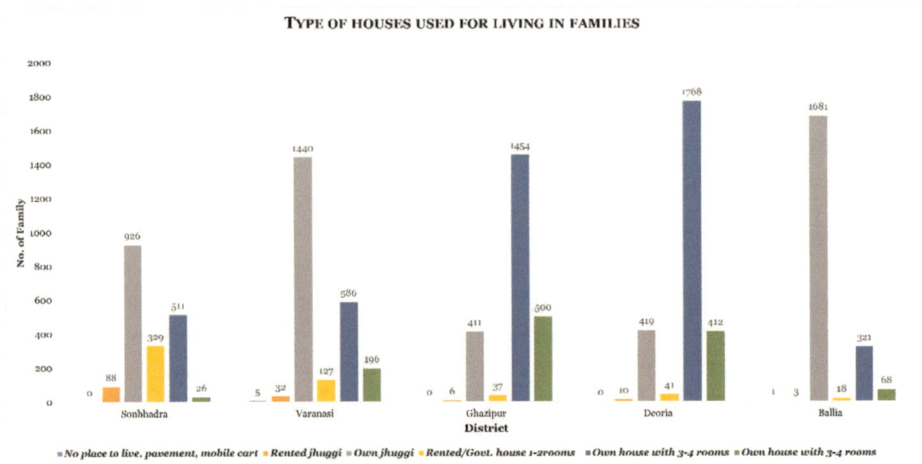

Figure 1.
Graph showing type of house used for living in each district.

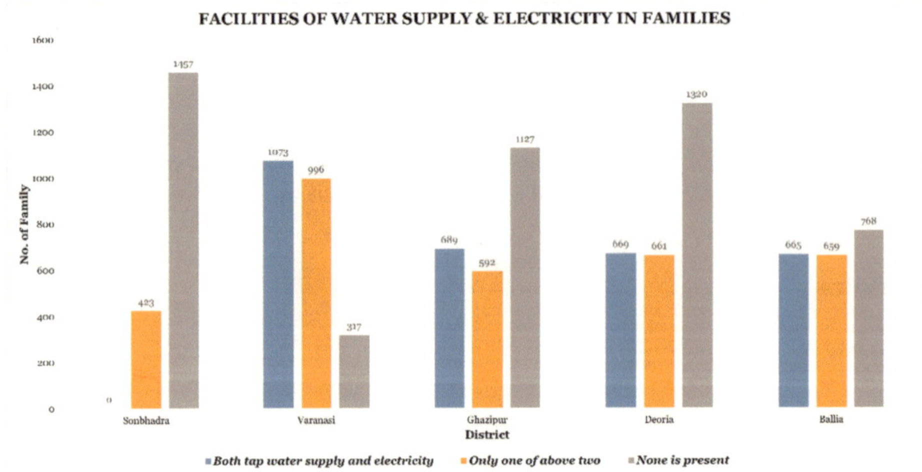

Figure 2.
Graph showing water and electricity facilities in each district.

Drinking water parameters		Number of families (%)	District wise distribution: number (%)				
			Sonbhadra	Varanasi	Ghazipur	Deoria	Ballia
Source	Tap water	188 (1.64%)	0	77 (3.2%)	97 (4%)	8 (0.3%)	6 (0.3%)
	Hand- pump	10,520 (92.15%)	1522 (80.95%)	2231 (93.5%)	2049 (85.1%)	2639 (99.5%)	2079 (99.6%)
	Boring	336 (2.94%)	0	67 (2.8%)	261 (10.8%)	2 (0.07%)	6 (0.3%)
	Well	364 (3.18%)	350 (18.6%)	11 (0.5%)	1 (0.04%)	1 (0.03%)	1 (0.05%)
	Pond	8 (0.07%)	8 (0.43%)	0	0	0	0

Drinking water parameters		Number of families (%)	District wise distribution: number (%)				
			Sonbhadra	Varanasi	Ghazipur	Deoria	Ballia
Location	Within house	6090 (53.34%)	12 (0.6%)	692 (29%)	1771 (73.5%)	2480 (93.6%)	1135 (54.3%)
	Neighboring area	4169 (36.51%)	1528 (81.3%)	1037 (43.5%)	629 (26.1%)	161 (6.08%)	814 (38.9%)
	Outside away	1157 (10.13%)	340 (18.1%)	657 (27.5%)	8 (0.33%)	9 (0.33%)	143 (6.8%)
Purification method	Bleaching Chlorinated	188 (1.65%)	0	77 (3.2%)	97 (4%)	8 (0.3%)	6 (0.3%)
	Alum mix	0	0	0	0	0	0
	Machine	0	0	0	0	0	0
	Boiling	0	0	0	0	0	0
	None	11,228 (98.35%)	1880 (100%)	2309 (96.8%)	2311 (96%)	2642 (99.7%)	2086 (99.7%)
Total		11,416	1880	2386	2408	2650	2092

Table 1.
Drinking water supply source, distance, and purification status.

from hand-pump followed by submersible boring (2.94%), well (3.18%), and tap water (1.64%). Pond is still source of drinking water for 0.07% of Sonbhadra families; they are devoid of tap water supply and submersible boring. The majority of drinking water supply is within house (53.34%) or in neighboring area (36.51%), but in Sonbhadra, it is mostly in neighboring area (81.3%) or outside away (18.1%). Almost all (98.35%) drinking water supply is untreated, the remaining 1.6% is bleaching chlorinated.

6. Sanitation

6.1 Facility of in-house sanitary latrine

Figure 3 presents family possession of sanitary latrines. Only 9.1% of ST families have sanitary latrine facility that is nil in Sonbhadra.

6.2 Sanitation status out of home

Figure 4 presents sanitation status outside the home of family. It shows that sanitation outside the home is satisfactory in only 46.95% of families.

7. Discussion

7.1 Housing

7.1.1 Type of house used for living

Majority of ST families live in either *jhuggis* (43.9%) or their own house with 1–2 rooms (45.5%), followed by own house with 3–4 rooms (10.5%). These findings

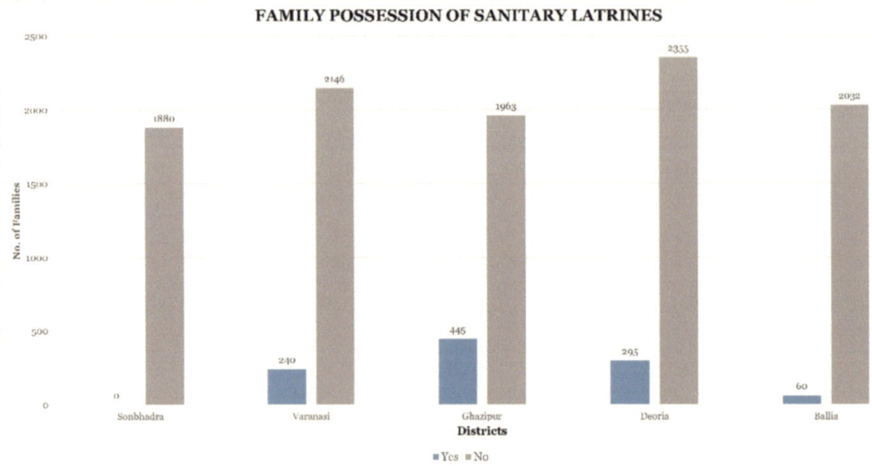

Figure 3.
Graph showing in-house sanitation facilities in each district.

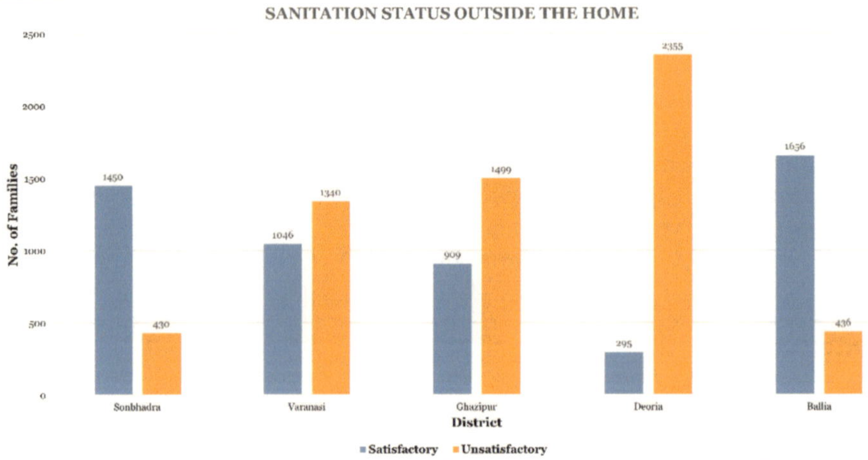

Figure 4.
Graph showing out house sanitation facilities in each district.

affirm the observation made by Singh that ST has very low level of physical conditions of living [1]. Sharma added that 20% of the tribal population has been uprooted and displaced within 50 years, they have lost their rights because of their political powerlessness, and the magnitude of land and number of displaced persons have been increasing since then [6].

Census 2011 data reveal that 40.62% of STs lived in good-condition houses and 6.2% lived in dilapidated houses compared to 53.1 and 5.35%, respectively, of the all-social groups [15]. In the present study, 54% of ST families in Sonbhadra lives in *jhuggis*, which is still high in comparison to 19.57% in general population of Sonbhadra. Similar is the status of houses of ST families in other districts (Ghazipur 23.3 and 6.22%, Ballia 80.18 and 8.66%, Deoria 15.84 and 8.54%, and Varanasi 61.7 and 2.15%). Present finding is in conformity with the finding of Jaiswal that more than 55% of *Kharwar* tribe stay in kuccha houses [12]. These findings reinforce the fact related to houses of *tharu* made of mud and brick, using and thatching wooden rod, traditionally, *tharus* house making system, agriculture system, cooking system was based on the nature of law, which is why the environmental balance never distorted in past. Culture of tribes is eco-friendly because of their deep relation with nature, which is also reflected in their living conditions [10].

Rai [10] reported that plans (such as *Awas Yojna*) are underway for the poor in rural areas but benefits are not transmitted to the beneficiaries due to several factors, including corruption. Some tribes do not have BPL cards in spite of their eligibility, and some have BPL cards but their name is not on the BPL list therefore they do not get benefit of such plans [10]. Present finding is also comparable to the observation presented by Bano and Ara [13] that most (92%) of *kharwar* tribe live in semicemented houses [13].

7.1.2 Locality of family residence

Most of the ST families are living either in rural localities (54%) or in *jhuggis*/slums (43.9%), devoid of basic facilities to live and earn. Most of them are slum dwellers due to hill terrain (Sonbhadra 53.9%), flood-affected area (Ballia 80.4%), and urban slum (Varanasi 61.7%) because of displacement and compulsion of temporary nature of livelihood while those of Ghazipur (81.7%) and Deoria (83.2%) in rural locality. Finding in the present study shows much higher percentage of locality of slum dwelt by families in study districts in comparison to that of general population as reported in census 2011. Finding in the present study shows much higher locality of *jhuggis*/slums dwelt by families in study districts in comparison to that of general population as reported in census 2011 (Sonbhadra 54% in study and 19.57% in census 2011, Ghazipur 23.3 and 6.22%, Ballia 80.18 and 8.66%, Deoria 15.84 and 8.54%, and Varanasi 61.7 and 2.15%, respectively).

7.2 Facilities of tap water supply and electricity

Only 27.12% of ST families have both tap water supplies and electricity and 43.7% have none of it. Sonbhadra families are without tap water supply and electricity connection is limited to 22.5% only, electricity connection in Sonbhadra at the present study is comparable to that of 20.94% in general population of Sonbhadra as reported in census 2011. Electricity supply observed as 45.6% in Varanasi, 24.9% in Ghazipur, 32.8% in Deoria, and 31.3% in Ballia families is comparable to those of 62.04, 20.15, 31.64, and 24.87%, respectively, to general population reported in census 2011; hence, almost all families are lagging behind the supplies of these essentials at present. Present finding is also in conformity with Economic Survey Report (2015–2016), which mentions that the disparity in terms of access to household amenities like tap water is sharp [11]. Facility of tap water is 89.5% in Himachal Pradesh, and 85.4% in Sikkim and Goa but only 27.3% in U.P. Present finding is also in conformity with observation of Jaiswal that more than 60% of tribal areas are not electrified [12]. Both clean energy and drinking water are included in SDG and countries, including India, are committed to achieve it, but there is no desirable progress in ST families [16].

7.3 Drinking water supply source, distance, and purification status

Most (92.15%) of ST families collect drinking water directly from hand-pump followed by submersible boring (2.94%), well (3.18%), and tap water (1.64%). Only

0.07% of them collect it from pond. Drinking water sources from ponds in 0.07% Sonbhadra families and no tap water supply and submersible boring is matter of public health concern. Sonbhadra ST families use hand pumps (80.95%) comparatively higher than general population (62.2%), but no tap water against 14.57% used by general population reported in census 2011. Findings in Ghazipur and Varanasi are also comparable to census 2011 general population. Findings in Ballia are lower (mainly tap water and boring) in comparison to census 2011 general population mentioning tap water (15.56%) and boring (0.31%) [15]. Hence, present finding reveals very low achievement of piped tap water supply in ST families in comparison to general population and they are more dependent on hand pump for drinking water.

In the present study, the majority of drinking water supply is either within house (53.34%) or in neighboring area (36.51%) on average but in Sonbhadra it is mostly in neighboring area (81.3%) or outside away (18.1%). Almost all (98.35%) drinking water supply is untreated, and only 1.6% is bleaching chlorinated due to direct collection from hand pump. The current results are consistent with census 2011 data, which note that the availability of drinking water paints a dismal image because just 19.72% of STs have a source inside their buildings, while 33.59% have one outside. In this aspect, the other group does better (46.6% and 17.6%). As the primary source of drinking water, hand pumps are used by both STs and all categories (33.5%) and 39.2% of STs, respectively. For all homes in social groups, treated tap water ranks as the second most accessible source (32%), while for self-taught people (19.1%), the most accessible source is untreated healthy water.

Almost all (98.35%) drinking water supply is untreated due to direct collection by hand pump. Although government is taking steps for piped water supply, ST family lacks it due to their scattered nature of remote placed *kaccha* houses and lack of proper attention from responsible authority. Untreated drinking water taken from uncovered wells and polluted ponds is an important public health problem.

Present finding is also in conformity with Economic Survey Report (2015–2016) related to drinking water supply mentioned earlier. Facility of drinking water within the premises is as high as 85.9% in Punjab; it is only 51.9% in U.P. The National Rural Drinking Water Programme launched a new project, supported by the World Bank, to provide a safe, 24 × 7 piped drinking water supply to 7.8 million rural people in four low-income states, namely Assam, Bihar, Uttar Pradesh, and Jharkhand, which have the least piped water supply and sanitation facilities [11]. Jaiswal also mentioned that half of the ST population lacks availability of pure drinking water [12].

7.4 Sanitation

7.4.1 Facility of in-house sanitary latrine

Present finding of possession of sanitary latrine of only 9.1% is comparable to the national average, which mentions that in India, an exceedingly high 77.4% of STs do not have latrine facility inside the premises as compared to 53.1% of all population. However, it is very low in comparison to Government of India 2013, which mentioned that only 46.9% of all households out of which 22.6% of ST households have latrine facility within the premises and 74.7% of ST households are still going for open defecation [2]. Present finding is in conformity with the Economic Survey, 2015–2016 report that presented that only 35.7% of population in U.P. had access to and coverage of latrine facilities, which was as high as 95% in Kerala and 91% in Mizoram. Achievement of sanitary latrine in present study is dismal in spite of the fact that the

Government of India launched *Swachh Bharat Mission* and sanitation coverage, which stood at 40.6% as per NSSO, has risen to around 48.8% (as of December 2015) [11].

As per census 2011, more than 72% of the rural population defecates in rural area, which is even more in ST population. Lack of sanitation facilities puts at risk not just public health and pollution but also security, particularly for vulnerable women, leading to severe crimes such as rape, sexual assault, and eve-teasing. Lack of sanitation services leads to pollution, health challenges, and, most importantly, security issues. Women should practice open defecation, which puts them more exposed to crime. Rape, sexual assault, or eve-teasing frequently occur in the dead of night, and the screams of anguish are never heard [15].

Sonbhadra ST family in the present study did not possess sanitary latrine in contrast to that of 25.83% in general population. All of Sonbhadra ST still go for open defecation in contrast to 74.18% in general population of Sonbhadra as reported in census 2011. Possession of sanitary latrine in the present study in Ghazipur (18.5%) is still lower in comparison to 21.89% in general population of Ghazipur reported in census 2011. Possession of sanitary latrine in the present study in Ballia (2.9%), Deoria (11.1%), and Varanasi (10.1%) is still very low in comparison to 26.88% in general population (26.98, 22.8, and 55.91% as reported in census 2011. Figure of open defecation by ST families in the present study (Sonbhadra 100%, Ballia 97.1%, Deoria 88.9%, Ghazipur 81.5%, and Varanasi 89.9%) is still higher than that of general population (74.18, 77.2, 77.75, 76.69, and 43.83%) as reported in census 2011 [15, 17].

7.4.2 Sanitation status out of home

Sanitation of neighboring surrounding is unsatisfactory in 53.05% of families. Census of India 2011 revealed "lack of basic sanitation and unhygienic living conditions, as around 70% of India's population (650 million) lives in rural and slum area." Present finding of poor sanitation in surrounding is associated with location of their residence predominantly in slum area/*jhuggi* (43.9%) and rural locality (54%) having unhygienic drainage and surface sanitation as well as their nonpossession of sanitary latrines (90.9%) leading to open defecation [15]. Jaiswal also reported that they do not have any proper sanitation facilities. Their knowledge of health and sanitation is very poor, they are poor at cleaning their own house [12].

8. Conclusion and recommendations

Most ST families still live in either *jhuggis* (43.9%) or own houses with 1–2 rooms (45.5%); they are living either in rural locality (54%) or in slum (43.9%), devoid of basic facilities to live and earn. Such poor conditions due to the terrain of hill and flood, displacement, and compulsion of temporary nature of livelihood; low standard locality is due to the effect of forced migration and urbanization. Although welfare plans such as subsidizing housing like *Lohia, Indira, Kashiram Awas Yojna. PM Yojna* exists for poor in rural areas, but tribes are not getting benefits; their housing condition continues to remain worse compared to previous census data and other social categories. Some tribes do not have BPL cards in spite of their eligibility; therefore, they do not get benefit of such plans.

Only 27.12% of ST families have both tap water supplies and electricity and 43.7% have none of it, facilities are much lower than general population; this disparity is more marked in Sonbhadra. In spite of the government's commitment to achieve both clean energy and drinking piped water as part of SDG, benefit is slow among ST families in eastern U.P. They have very low (1.64%) achievement of piped tap water supply in comparison to general population and they are more dependent on hand pump (92.15%) for drinking water. Digging of small pit in the land locally called "*kuhaad,*" which collects water from drains and spring in it, along with pond water (0.07%) in Sonbhadra district pose their exposure to contaminated water not only with germs but also excess iron, fluoride, and heavy metals, leading to deformity and increased mortality [18].

Access to drinking water source is only 53.34% within home and they are dependent on neighbor as high as 81.3% in Sonbhadra, which remain distant compared to general population in spite of drinking water mission. Almost all (98.35%) drinking water supply is untreated due to direct collection by hand pump. It is due to the scattered nature of remote placed *kaccha* houses and lack of proper attention from responsible authority. Necessary corrective measures are needed to provide pure potable water to address highly prevalent water-born public health problem.

As high as 77.4% of STs do not have latrine facility inside the premises as compared to 53.1% of all India and 64.3% of U.P. general population, and achievement of sanitary latrine is dismal in spite of the *Swachh Bharat Mission*. Figures of open defecation by ST families are still higher than general population reported in census 2011 and it is a matter of great concern, a problem throughout the Indian subcontinent. It poses not only public health and pollution problems but also security problems especially for vulnerable women leading to serious crimes such as rape, sexual assault, and eve-teasing. Provision of adequate sanitation facilities will lead to improvement not only overall status but also reduction in serious crimes against the weaker society, which is still very high among these communities. Sanitation in their neighboring area is unsatisfactory in 53.05% of families mainly due to unhygienic drainage and surface sanitation in slum area. The construction of drainage system, village sanitation infrastructure, personal toilets, and the environmental measures to control mosquito breeding should be included in the MGNREGA scheme and completed on priority basis in Scheduled Areas [15, 19, 20].

Due to low education and economic factors, tribes are victims of inequality, exploitation, and oppression. Tribes of backward eastern U.P. are living in conditions of deprivation; their economic condition due to subsistence low level of economy and standard of living are very low, as most of them do not have land, assets, and education. Protective developmental measures have not yielded any remarkable impact on tribal development; special budget provision remains unutilized largely. Although rich limestone hills in Sonbhadra have given establishment of cement and other allied factories and giant thermal plants, native tribes are not getting desired benefit. The low representation of tribes to the total population often excludes them from development processes hence their adequate political representation is required for their uplift and empowerment. There is an urgent need for robust institutions to not only bridge wide gaps between ST and general population in rapidly changing socio-economic conditions but also strengthen social inclusion.

Acknowledgements

I would like to acknowledge the IOE Banaras Hindu University and Indian Council of Social Science Research, New Delhi, and express my special gratitude for collaboration. I wish to thank various people for their contribution to this project.

Funding

The author received financial support from Indian Council of Social Research, New Delhi for the research of this article.

Declaration

The author declared no potential conflicts of interest with respect to the research, authorship, and/or publication of this article.

Author details

Poonam Singh Kharwar[1]*, Devesh Kumar[2], Abhishek Kumar[3] and Abhinav Kumar[4]

1 Banaras Hindu University, Varanasi, India

2 Banaras Locomotive Works - Indian Railway, Varanasi, India

3 Indian Institute of Technology, Kharagpur, India

4 All India Institute of Medical Science, Bhubaneshwar, India

*Address all correspondence to: poonam.kharwar3@gmail.com

References

[1] Singh AK. Endangered Tribals in India: Booby trap of development. Social Change. 1997;**27**

[2] Government of India. Statistical Profiles of Scheduled Tribes in India 2013, Ministry of Tribal Affairs. Available from: http://www.tribal.nic.in/ST/StatisticalProfileof STs2013.pdf

[3] Global Data Lab. Construction of the SHDI and SGDI Indices. 2021. Available from: https://hdi.globaldatalab.org

[4] Rai RK. Tribes of Uttar Pradesh: Brief introduction. IJAR. 2018;**4**(1):1143-1149

[5] Belshaw CS. Development: The contribution of anthropology. International Social Science Journal. 1972

[6] Sharma BD. Globalisation: The Tribal Encounter. New Delhi: Har Anand; 1995

[7] Mehta. Dynamics of Tribal Development. Delhi: Anmol Publications; 2000

[8] Mondal A, Mete J. Tribal development in India: Educational perspectives. Indian Streams Research Journal. 2012;**2**(11)

[9] Desmukh BA. Tribal development approaches: A theoretical prospective. Tribal Research Bulletin, TRTI, Pune. 2003

[10] Rai RK. Socio-political aspects of tribal community: A case study of five districts (Ballia, Mau, Gorakhpur, Deoria, Maharajganj) of Purvanchal in Uttar Pradesh. International Journal of Current Research. 2017;**9**(1): 45182-45188

[11] Economic Survey 2015–16. Delhi: Young Global Publications

[12] Jaiswal A. Kharwar: A Dynamics of Change. New Delhi: Alfa Publications; 2017

[13] Bano MJ, Ara A. A study on socio-cultural life of kharwar tribe in Sonbhadra. IOSR Journal of Humanities and Social Sciences. 2018;**23**(2):16-22

[14] Aggarwal OP, Bhasin SK, Sharma AK, Chhabra P, Aggarwal K, Rajoura OP. A new instrument (scale) for measuring the socioeconomic status of a family. Indian Journal of Community Medicine. 2005;**30**(4)

[15] Census. Registrar General of India, Demographic Status of Scheduled Tribe Population in India, New Delhi; 2011

[16] Govt. of India. An Overview of the Sustainable Development Goals. New Delhi: NITI Aayog; 2016. Available from: http://niti.gov.in/content/overview-sustainable-development-goals

[17] Agriculture Census. Department of Agriculture and Farmers Welfare, Ministry of Agriculture and Farmers Welfare, New Delhi. 2011. Available from: agcensus.nic.in/document/agcensus2010/allindia201011H.pdf

[18] Government of UP. Statistical Diary Uttar Pradesh, Economics and Statistics Division, State Planning Institute Planning Department, Uttar Pradesh, Lucknow; 2017

[19] Government of India. Handbook on Social Welfare Statistics, Ministry of Social Justice and Empowerment, New Delhi; 2018

[20] Government of India. Report of the Advisory Committee on the revision of list of SC and STs, Department of Social Security, New Delhi. 1965. Available from: https://tribal.nic.in/writereaddata/AnnualReport/LokurCommittee Report.pdf

Section 2

Perspectives and New Technologies

Chapter 4

Perspective Chapter: From Data to Design – Leveraging Façade Sensors for Intelligent Architecture

Mubarak Reme Ibrahim

Abstract

This chapter explores the fascinating domain of leveraging façade sensors for intelligent architecture, focusing on the seamless transition from data to design. This study will delve into the integration of advanced sensor technologies within building façades to collect valuable data that inform the architectural design process. This chapter investigates how these sensors provide real-time information on various aspects, such as environmental conditions, occupancy and energy usage, enabling architects to design responsive, sustainable and occupant-centric buildings. Architects can improve building performance, optimise user experience and shape the future of intelligent architecture by harnessing the capabilities of façade sensors.

Keywords: sensors, façades, integration, monitoring, performance, real-time data, intelligent architecture, occupancy sensor

1. Introduction

The built environment has been significantly affected by the incorporation of technology into architecture. Developments in sensor technologies have paved the way for intelligent architecture. The design, construction and operation of buildings can be completely transformed by the use of sensors [1]. By integrating these sensors, architects can gain real-time insights into a building's surroundings, empowering them to design façades that respond intelligently to climate, natural ventilation, solar heat gain and daylighting. As a result, buildings have become more sustainable and energy efficient, decreasing their reliance on artificial lighting, heating and ventilation systems [2]. To design spaces that satisfy the needs of their occupants, architects can use façade sensors to monitor occupancy patterns. These data assist architects in designing spaces that meet their requirements, such as optimising low-occupancy spaces to save energy while still providing sufficient social and seating areas for high occupancy. By identifying unauthorised access or unusual movement patterns, occupancy sensor data can also help improve building security [3].

Façade sensors also play a significant role in monitoring and improving energy utilisation in buildings by coordinating energy use sensors into the façade, and architects can monitor and analyse the use of power, water and different utilities, empowering them to make informed design decisions that focus on energy efficiency [4]. These data can be used as a guide for design interventions such as better insulation, appliances that use less energy and renewable energy systems. It can also be used to show building users' real-time energy consumption data to encourage environmentally friendly behaviour. The capability of façade sensors is not only to collect data but also to translate these data into practical recommendations. Architects can now use real-time data to make informed design decisions thanks to advancements in data analytics and visualisation methods. This iterative design process permits architects to enhance their designs for numerous parameters simultaneously, guaranteeing a balance between aesthetic, function and performance considerations.

The integration of façade sensors into intelligent architecture has far-reaching implications beyond the planning stage of a project. By enhancing the user experience and contributing to a more sustainable and occupant-centric built environment, buildings that use façade sensors can shape the future of intelligent architecture. Façade sensors help reduce a building's impact on the environment while also improving user health and safety, productivity and satisfaction. For example, buildings can create spaces that adjust to individual inclinations, advancing occupant comfort and satisfaction by tailoring the indoor climate based on real-time data. The seamless change from data to design is reforming the field of architecture by providing architects with a deeper understanding of different aspects of design, such as site conditions, environmental factors, building performance and user behaviour. Because of this change in practise, architects are now able to make design decisions based on evidence, which enables them to come up with solutions that are more responsive and appropriate to the context. The capacity to enhance building performance is one of the primary advantages of the seamless transition from data to design. The seamless integration of data into the design process also enables architects to evaluate design alternatives more effectively. Before construction begins, data-driven simulations and modelling can help architects evaluate the performance of various design options, thereby minimising errors and avoiding costly modifications during construction. In addition, the effortless transition from data to design encourages interdisciplinarity and collaboration. For holistic and integrated design solutions, architects can work with builders, environmental consultants and other relevant professionals to collect and analyse relevant data [5].

However, there are obstacles in achieving a seamless transition from data to design. One of the major questions is the shear volume and intricacy of the information accessible. Architects must be equipped with the necessary skills and tools to collect, analyse and interpret data effectively. Realising the potential of data-driven design requires embracing parametric modelling, data visualisation, and computational design methods. The incorporation of data into the design process itself is yet another disadvantage. To achieve a balance between artistic vision and empirical evidence, architects must adopt a mindset that views data as an essential resource [6]. This chapter explores the seamless application of façade sensors in intelligent architecture. This study focuses on the collection and analysis of valuable data as it explores the incorporation of cutting-edge sensor technologies into building façades. This chapter features the pragmatic uses of façade sensors, such as monitoring climate data, predicting user behaviour and optimising energy consumption. Additionally, it emphasises the advantages of user-centred design, flexible spaces, responsive design and the incorporation of façade sensors into intelligent architecture. This chapter accentuates the enhancement of building performance through sensor-driven tasks, propelling sustainability by designing green buildings and the inventive capability of occupant-centric buildings, as well as enhancing health and safety.

2. Overview of façade sensors

In architecture, façade sensors play a key role in the development of smart and sustainable buildings. These sensors are integrated into the exterior surfaces of a building's façade to monitor and respond to various environmental factors, occupancy and energy usage. They offer rich data and insights that architects, builders and building operators can use to optimise the building's performance, comfort and energy efficiency.

2.1 Definition and type of façade sensors

2.1.1 Definition façade sensors

Façade sensors, also referred to as building envelope sensors or climate sensors, are instruments strategically positioned on the exterior surface of a building's façade to measure and monitor several environmental parameters [7]. These parameters include temperature, humidity, wind speed, solar radiation, air quality and noise levels.

Façade sensors are specialised devices placed on the outer surface of buildings to retrieve data on environmental conditions such as temperature, humidity, light levels and wind speed, which are crucial for improving energy consumption and comfort [8].

Façade sensors entail various sensing technologies installed in the building envelope. These sensors enable real-time monitoring and assessment of external factors, enabling responsive and energy-efficient building operations [9].

Façade sensors are tools incorporated on the external faces of buildings to detect environmental parameters. They offer data that supports smart building management systems, allowing for dynamic modifications in lighting, heating and ventilation to conserve energy and improve comfort [10].

Façade sensors include multiple sensors and data gathering instruments integrated into a building's façade. These devices continuously monitor parameters such as solar radiation, air quality and temperature. The retrieved data are used to improve energy utilisation and form sustainable and responsive building environments [11].

2.1.2 Types of façade sensors

2.1.2.1 Temperature sensors

Temperature sensors are important for contemporary architecture because they provide valuable data that helps architects and building operators create efficient, comfortable and sustainable environments. These sensors are integral components of smart buildings, enabling real-time monitoring and control of indoor and outdoor temperatures. Their application in architecture spans several phases, from initial design to post-occupancy evaluation. In the early stages of architectural design, temperature sensors facilitate building energy simulations and modelling. They assist architects in assessing how sunlight and outdoor temperature variations affect the building's thermal comfort. This insight enables the design of passive heating and cooling systems, such as building orientation, shading and natural ventilation, to reduce the dependence on energy-consuming heating, ventilation and air conditioning (HVAC) systems. After a building is completed, temperature sensors are installed indoors to sustain thermal comfort for the occupants. These sensors monitor indoor temperatures, ensuring that they remain within a designed range for different locations of the building. Architects can create responsive and zoned climate control, allowing personalised comfort for occupants while optimising energy consumption by linking temperature data to HVAC systems [12].

Temperature sensors provide real-time data that can be incorporated into building management systems. This data allow for the dynamic regulation of heating, cooling and ventilation systems based on existing situations. For example, if a class room experiences a sudden increase in occupancy, temperature sensors can signal the HVAC system to adapt and sustain an optimised environment. When combined with occupancy sensors, temperature sensors facilitate smart climate control systems that adjust to the number of people in an area. Unused spaces can be conditioned to a reduced level, reducing energy use while ensuring comfort when occupants are around. Architects are gradually integrating temperature sensors into adaptive façades and building envelopes. These sensors deliver data to manage dynamic shading devices, louvres and phase-changing materials that respond to outdoor weather conditions. This approach improves energy efficiency and reduces the building's environmental impact. When the building is in use, temperature sensors continue to contribute to its performance evaluation. Architects can identify areas for improvement and fine-tune the building's systems to achieve superior performance and comfort by evaluating temperature data with a combination of energy consumption data. In larger construction projects, temperature sensors play a role in smart grid integration. Buildings equipped with temperature sensors can participate in demand response programmes, regulating their energy utilisation during peak periods, thus enabling grid stability [13].

2.1.2.2 Humidity sensors

Humidity sensors gauge the amount of moisture in the air, providing data that assists architects and facility managers in optimising HVAC systems, preventing moisture-related issues and improving the general health and safety of occupants. Humidity sensors are crucial for sustaining optimal indoor air quality and comfort. Architects can guarantee that indoor areas remain within the planned relative humidity level (typically 30–60%) [12]. Appropriate humidity regulation hinders the growth of mould and mildew, minimises the risk of respiratory issues, and promotes a better indoor environment by continuously controlling humidity levels. In areas with fluctuating weather, humidity sensors help avoid condensation problems on windows and walls. By providing real-time humidity data, these sensors enable architects to design proper insulation ventilation systems. Humidity sensors are influential in improving HVAC systems, mainly in humid environments. When incorporated with building automation systems, these sensors allow smart control of humidification and dehumidification processes and reduce energy waste [14].

Humidity levels can, to a large extent influence, the durability of building materials such as timber, steel and paint. Architects can prolong the service life of these materials, thereby reducing maintenance budgets and preserving the building's aesthetics by maintaining suitable humidity levels. Humidity sensors are essential in critical spaces of the building, such as bathrooms, kitchens and basements, where moisture levels tend to vary. Early identification of high humidity in these areas can enable proper ventilation and humidity control measures, mitigating potential dampness and growth of biological deterioration agents. In construction projects involving indoor agriculture or greenhouses, humidity sensors play a critical role in improving the growth environment for plants [12]. These sensors help maintain the appropriate level of humidity to support plant growth and prevent pest diseases. Humidity sensors allow for data-driven design and post-occupancy evaluations. Building designers can identify areas for improvement, optimise system performance and ensure that the building meets its performance objectives by analysing humidity data with other environmental parameters.

2.1.2.3 Light sensors in the architecture

Light sensors, also referred to as photodetectors, are instrumental devices in modern architecture, that enable accurate monitoring of artificial lighting systems and optimise energy efficiency. These sensors create green and human-centric built environments, by providing real-time data on natural and artificial light conditions. Light sensors are an integral element of energy-efficient lighting systems in architecture [15]. By measuring the intensity of natural light in a space, light sensors enable the automatic regulation of artificial lighting [16]. When daylight levels are satisfactory, artificial lighting dims or turns off, thereby minimising electricity usage immensely. This dynamic lighting control system not only lowers energy consumption but also reduces greenhouse gas emissions. Integrating light sensors into building design allows effective daylight harvesting, an approach that uses natural light as the main source of lighting. Light sensors can detect the varying daylight levels and match with lighting regulators to sustain a continuous lighting condition. Buildings can attain higher energy efficiency and Leadership in Energy and Environmental Design (LEED) certification by reducing the use of natural light [17]. Light sensors contribute to improving occupant health and safety by considering the human circadian rhythm. These sensors can control artificial lighting intensity and colour temperature during the day to align with sunlight's varying qualities. Buildings can positively impact occupants' sleep behaviours, productivity, and well-being by promoting a circadian lighting environment [18].

Light sensors are often integrated with occupancy sensors to produce occupant-responsive lighting systems. When occupants are identified in a specific location, the light sensors evaluate the available sunlight and alter the artificial lighting levels. Thus, energy is preserved by providing lighting only when needed, thereby minimising running costs [19]. Combining light sensors with adaptive façades and building envelopes enhances the sustainability of architecture. These sensors detect incoming artificial lighting and alter shading devices, glass tinting or dynamic elements on the façade to improve sunlight diffusion and thermal comfort [20]. The adaptive nature of the façades reduces the building's energy demands, forming a more environmentally friendly design. Light sensors allow architects to assess the effectiveness of lighting design approaches. Architects can fine-tune their design and improve the building's performance by collecting data on light levels, usage behaviours and occupant feedback [21, 22].

2.1.2.4 Air quality sensors

Air quality sensors are indispensable devices in smart architecture, allowing precise monitoring and control of indoor air contaminants. These sensors are crucial for creating healthy and comfortable indoor environments while contributing to energy efficiency and sustainability. Air quality sensors allow architects to consistently control indoor air pollutants, such as volatile organic compounds (VOCs), particulate matter, carbon dioxide (CO_2) and formaldehyde [23]. These sensors allow timely responses to variations in contaminant levels, ensuring that occupants breathe clean and safe air by offering real-time data on air quality. Air quality sensors are combined with ventilation systems to facilitate demand-based ventilation. When pollutant levels are above satisfactory thresholds, the sensors activate increased ventilation flow rates to eliminate toxins and improve indoor air quality [24]. This adaptive ventilation strategy not only improves occupant comfort but also reduces electricity consumption by avoiding unnecessary ventilation. Maintaining good indoor air quality is critical for the welfare and productivity of building occupants. Air quality sensors facilitate the identification of harmful contaminants that can cause respiratory diseases. Architects can contribute to a healthier and more productive indoor environment by proactively managing indoor air quality.

2.1.2.5 Solar radiation sensors

Solar radiation sensors, also referred to as solar pyranometers, are vital tools in modern architecture for evaluating solar energy availability and improving sustainable building design. These sensors allow architects to harness sunlight to enhance the energy efficiency of buildings. Solar radiation sensors play a crucial role in Site analysis during the early stages of architectural design [25]. By measuring solar irradiance, architects can evaluate solar access and detect areas with improved sunlight. These data help in determining the orientation of the building to reduce natural light consumption and eliminate the need for artificial lighting during the day. Architects can design passive solar systems that use solar energy for heating and lighting. By strategically placing windows, roof lights and shading instruments based on solar radiation data, buildings can achieve improved thermal comfort and energy efficiency [26]. Passive solar design reduces dependence on artificial heating and cooling systems, leading to energy savings and reducing carbon footprint.

For buildings that integrate photovoltaic cells (PV) systems, solar radiation sensors are essential for system improvement. By controlling solar irradiance levels, architects can determine the most suitable positions and angles for PV panels to increase energy generation. This improvement ensures that PV systems produce the highest electricity output from existing sunlight. Solar radiation sensors are incorporated into building energy simulation tools to model solar gains and evaluate building performance. Using solar radiation data, architects can simulate a building's energy performance, forecast cooling and heating demands, and assess passive solar design approaches [27]. This assessment helps in making informed design decisions to design energy-saving buildings. Solar radiation sensors contribute to dynamic façade design. These sensors can identify incoming solar radiation and activate responsive shading systems. The dynamic façade adapts to varying solar conditions, maintaining a comfortable indoor environment and lowering cooling needs.

2.1.2.6 Occupancy sensors

Occupancy sensors, also referred to as motion detectors, identify the movement of occupants in an area to allow accurate monitoring of lighting, heating, ventilation and air conditioning (HVAC) systems. Occupancy sensors are valuable in sustainable lighting control in architecture because they identifying the occupants in a space and activate automatic on/off of lighting systems [28]. Consequently, electricity usage is reduced by eliminating the need for illumination in vacant spaces. Combining occupancy sensors with HVAC systems allows for adaptive climate control in buildings.

When people are identified in an area, the sensors can alter the heating or cooling levels to maintain thermal comfort. In vacant spaces, the ventilation system can work at lower levels to save energy. Occupancy sensors enable personalised user experience in building spaces [29]. The sensors can adapt lighting, ventilation and other environmental parameters to satisfy user needs, which improves occupant indoor comfort.

These sensors are usually integrated with light sensors in a daylight harvesting system to regulate artificial lighting settings based on the available sunlight which will enable energy saving by lowering artificial lighting when sunlight is adequate. They also help in enhancing building security and safety. In vacant spaces, sensors can initiate lighting or security systems. When integrated with a building automation system, data from occupancy sensors aid smart and energy-saving building operation, which contributes to a green built environment. Architects use occupancy sensor data to inform data-based design, improve space allocation, identify opportunities for energy saving and comfort optimisation by analysing occupant behaviours.

2.1.2.7 Structural health monitoring sensors

Structural health monitoring (SHM) sensors allow consistent monitoring and surveillance of buildings. These ensure the safety, reliability and maintenance of the building. They facilitate real-time and consistent monitoring of the structural integrity of buildings. They are strategically installed in structural elements, such as slabs, beams, columns and foundations, to determine shear stress, deflection, vibration and uneven settlement. SHM sensors help in the timely identification of structural decay. Any changes in the behaviour of a building, such as differential settlement or sudden shock, can indicate potential issues; they provide a warning signal of potential structural problems for preventive maintenance [30]. With SHM sensors, architects can develop performance-driven design and appraisal methodologies. Instead of relying on design codes, architects can use real-time data from sensors to validate design assumptions and improve structural stability. This method results in cost-effective designs.

The application of micro-electro-mechanical systems (MEMS) sensors in the diagnosis of seismic capacity for historic structural glass systems indicates a new advancement in structural engineering. These sensors offer real-time monitoring capabilities, enabling continuous data capture of structural vibrations and responses. This information supports structural engineers in understanding the dynamic behaviour of these glass structures, mainly under seismic loads [31]. MEMS sensors, characterised by their high precision and sensitivity, can detect subtle structural movements and potential problems. In addition, MEMS sensors are important for evaluating the mechanical performance of glass façades subjected to seismic loads. These sensors monitor vibrations, strain and deformation in real time, providing invaluable data for structural engineers to analyse structural integrity during earthquakes. By allowing detailed data capture and assessment, MEMS sensors support informed decision-making for maintenance, improving the safety and resilience of glass façades in seismic-prone regions [32].

2.2 Integration of advanced sensor technologies into building façades

In the quest for smart architecture, the incorporation of smart sensor technologies into building façades has emerged as a promising strategy. These smart sensor technologies, ranging from occupancy sensors to light detectors, are critical to improving user experience, improving building performance and reducing carbon footprints. Photodetectors are important when incorporated into building façades for daylight harvesting. They gauge the quantity of sunlight in buildings and regulate artificial lighting settings. Through this adaptive lighting control, electricity usage in buildings can be reduced by using available natural sunlight. Sunlight harvesting not only conserves energy but also contributes to occupant health by providing natural lighting environment. Incorporating humidity detectors in building façades enables accurate climate control. These sensors monitor indoor humidity, allowing HVAC systems to regulate heating and ventilation by optimising thermal comfort in real time, which facilitates energy savings without detriment to user satisfaction [33]. Moreover, they help prevent condensation, ensuring a healthier indoor environment. They are also instrumental in creating energy-efficient spaces within building façades. These sensors detect the presence or absence of occupants in rooms and trigger the lighting and HVAC systems accordingly. By turning off lights and reducing heating or cooling in unoccupied areas, buildings can achieve substantial energy savings.

Solar radiation sensors, also called pyranometers, are important the design of green building façades. They gauge solar irradiance and support architects in determining building orientation and shading design [34]. Buildings can reduce dependence on mechanical lighting and ventilation systems, leading to less energy utilisation and environmental impact by leveraging sunlight. Incorporating air quality detectors in building façades aids in the realisation of a healthy indoor environment. These sensors detect volatile organic compounds (VOCs) and carbon dioxide (CO_2), providing sensible ventilation control [12]. Maintaining acceptable indoor air quality improves occupants' mental health and productivity, making it a critical consideration in contemporary architecture. Integrating proximity sensors in building façades allows touchless interaction in building areas. They sense hand gestures and enable touchless control of doors, elevators and windows. Touchless control encourages hygiene, reduces the spread of diseases and addresses post-COVID design considerations. The incorporation of SHM detectors in building façades, guarantees the safety and longevity of buildings by monitoring stresses and vibrations to assess structural integrity. The timely identification of structural problems through SHM detectors allows for preventive maintenance and disaster management [31].

2.3 Collection and analysis of sensor data

In the field of smart architecture, the collection and analysis of sensor data in building façades have become important for creating energy-saving, user-centric and resilient buildings. Smart sensor technologies installed in building façades enable real-time data gathering, giving architects valuable insight into numerous areas of building performance. Environmental sensor networks are embedded in building façades to offer real-time surveillance of environmental parameters [33]. They monitor temperature, humidity, air quality and solar radiation, assisting architects in understanding the building's response to environmental conditions. The data captured helps in enhancing indoor climate control systems, optimising occupant comfort and saving energy. Light sensors, installed in building façades, allow daylight harvesting through the measurement of sunlight intensities [35]. This data-driven method enables for adaptive lighting adjustment, where mechanical lighting is controlled based on available sunlight. Sunlight harvesting also improves the health of the occupants by creating a dynamic lighting environment.

Occupancy sensors, a vital component of building façades, detect the presence or absence of occupants in various spaces [3]. The data captured by these detectors enable energy-efficient building operations by activating lighting and climate control systems based on real-time occupancy behaviours. This automated system enhances resource use and agrees with green building agenda. SHM sensors are embedded in building façades to monitor their structural integrity [32]. Data from SHM detectors enable architects to perform early maintenance to ensure the safety and resilience of the building. The valuable data retrieved from several detectors in building façades allow designers to embrace data-informed design solutions and decision-making [36]. Analysing the retrieved data enables for evidence-based design adjustments. Data-driven decisions lead to better space allocation, improved energy efficiency and improved user satisfaction. Incorporating data retrieved from building façade detectors into smart building systems improves building performance. Data integration enables dynamic building operations, where different systems work together to enhance energy utilisation, user experience and sustainability. Smart building sensors leverage big data to create an adaptive and user-centric building environment.

3. Real-time information on architectural design

3.1 Environmental condition sensors

In the search for sustainable and climate-responsive architecture, monitoring and analysing climate data play a significant role in shaping state-of-the-art façade. Façades, as the outer skin of buildings, act as the interface between the outdoor and indoor environments. Architects can design façades that adapt to dynamic weather by integrating climate data into the design processes. Monitoring climate data offers a better understanding of local weather patterns, temperature and radiation levels. This information aids architects in choosing suitable façade materials, shapes and shading to enhance building performance. Climate data-driven design approaches enable the development of passive façades that respond to environmental conditions, thereby reducing the building's reliance on artificial heating, ventilation and lighting. Analysing climate data is important for optimising thermal comfort in façade design. Understanding temperature variations and wind patterns helps architects implement passive design measures that control internal temperature [37]. Double-skin façades, thermal mass incorporation and natural ventilation can drastically impact the internal thermal environment and safeguard user health. Climate data inform sunlight harvesting agendas and solar gain control in façade design. Architects can determine optimum glazing ratios, orientation and shading to reduce sunlight while minimising undesirable solar heat gain by analysing solar radiation data. This approach optimises visual comfort and supports circadian rhythms. Integrating climate data into façade design allows climate adaptation measures. Architects can design façades that resist heavy rains, lateral forces and heat by understanding the climate conditions [38]. Climate adaptive façades contribute to the long-term durability and functionality of buildings in dynamic climates.

Monitoring and analysing climate data are key in considering the embodied energy of façade materials. Studying the environmental impacts of material selection aids architects in choosing green materials [39]. Climate-conscious material selection helps reduce the greenhouse gases in buildings. Climate data plays a critical role in forecasting and measuring building energy performance. Energy simulation platforms, integrated with climate data, allow architects to assess the energy efficiency of different façade design strategies [40]. This analysis allows for iterative design processes, leading to energy-efficient façades that align with sustainable goals. Monitoring and analysis of climate data after construction aid data-driven façade improvement. User experience helps architects evaluate the actual performance of the façade design. Real-world data aids in detecting spaces for optimising design, ensuring that the façade performs as intended during its life cycle.

Incorporating weather patterns into façade design decisions has emerged as a transformative approach. Architects can design façades that adjust to change environmental conditions, and enhance occupant comfort by harnessing. Integrating weather patterns into façade design includes using climate data to inform design strategies. By evaluating previous weather data and climate forecasts, architects can obtain insights into temperature variation, radiation level and wind direction [41]. This data-driven methodology allows the design of façades that respond to environmental conditions, facilitating passive design that reduces dependence on artificial heating, ventilation and lighting. Integration of weather patterns into façade design improves thermal comfort for building occupants. Understanding temperature variations and wind patterns allows architects to optimise the façade's thermal performance [37]. Features such as advanced insulation, glazing with appropriate solar heat gain coefficients, and passive solar design principles contribute to a comfortable indoor environment, thereby reducing energy consumption and improving occupant well-being.

Climate data empower architects to harness sunlight and control solar gain. By analysing solar radiation data, architects can improve façade fenestration, shading systems and light redirection elements. Amplifying sunlight while minimising heat gain reduces the building's energy load and creates visually comfortable and healthy spaces. Weather-based façade design contributes to climate adaptation. Integrating weather patterns into façade design aligns with sustainability goals. Climate-driven design includes selecting materials with less environmental impact that lower energy consumption [38]. Such considerations foster environmentally friendly architecture and reduce the building carbon footprint. Monitoring and analysing climate data enable energy improvement in façade design. Building performance simulations, combined with climate data, allow architects to assess the energy efficiency of different façade [4]. Monitoring weather patterns after construction aids in data-driven façade improvement. Real-world performance feedback informs architects of the façade's actual response to the climate. This data-driven approach allows for fine-tuning, ensuring that the façade remains climate-responsive throughout its lifespan.

3.2 Occupancy sensors

Architects are increasingly integrating technology to track user behaviour in buildings. This method seeks to enhance creating user-centric buildings. Designers can understand how occupants use the building façade, enabling for bespoke design that responds to change preferences by leveraging data analytics. Monitoring user behaviour enables architects to understand how occupants interact with the façade. Designers can identify patterns of user movement by analysing data from motion sensors and occupancy detectors [42]. This data-driven strategy allows architects to design façades that cater to specific user preferences. Architects can anticipate user preferences and adjust the façade's features. For instance, the façade could automatically adjust shading, ventilation or lighting based on user comfort needs, resulting in a user-friendly building. Tracking occupant behaviour contributes to optimising thermal comfort. Architects can enhance façade design to provide adequate sunlight and regulate the indoor environment by tracking occupant movements. Real-time data feedback informs design decisions, leading to efficient sunlight use and reducing the need for artificial lighting.

Predicting occupant behaviour promotes architectural innovation. Continuous data gathering and analysis enable architects to assess the efficiency of design [43]. This approach leads to innovative façade designs that evolve with occupant preferences. Design decisions are driven by the goal of designing buildings that ensure the health and safety of occupants. This approach promotes a stronger link between the building and its users, resulting in more functional spaces. Data-driven insights into occupant behaviour support building maintenance. Facility managers can enhance space allocation and manage resources by understanding user preferences.

Designing façades based on user behaviour includes the incorporation of sensors in buildings. Motion detectors and thermal cameras are used to gather data on occupant's behaviour [28]. These sensors offer real-time and past data, which architects can leverage to enhance façade design for occupant comfort. Architects can design façades that respond in real-time to the changing preferences of building users by analysing user activity data. Designing façades based on occupancy behaviour improves user experience. Understanding how occupants use spaces aids architects in designing intuitive layouts [44]. Architects can fine-tune building systems to align with actual needs by monitoring occupant movement. This approach supports environmentally responsible design. Data on occupancy patterns aid facility management. Facility managers can use the data to improve HVAC systems, and manage space allocation [45].

3.3 Energy usage sensors

Monitoring energy usage in façade design includes the incorporation of sensor technology. Smart metres, occupancy sensors and environmental detectors aid real-time data collection on energy usage, indoor environments and lighting conditions [28]. This information informs architects about building performance. They are vital in enhancing energy savings through façade design. Architects can detect energy-intensive spaces and energy-saving opportunities by analysing sensor data. This approach facilitates evidence-based design strategies, resulting in façades that minimise operational costs. Façade design can integrate passive strategies such as thermal mass, shading devices and natural ventilation to regulate indoor airflow and optimise energy consumption. Monitoring energy usage through façade design aligns with green building practises. Implementing energy-efficient façades is essential for achieving global sustainability goals [44]. Façade design can incorporate passive strategies to enhance sustainability. Passive design harnesses natural resources, enhancing daylight and thermal conditions to reduce the need for HVAC.

Façade design can incorporate energy-saving glazing. By using high-performance glazing with low U-values and effective insulation materials, architects can greatly reduce energy loss [46]. Improved glazing contributes to better thermal comfort, thus creating a sustainable building. Dynamic façades offer an innovative approach to energy saving. By incorporating sensors, façades can adjust to varying environmental conditions and user preferences. Switchable glazing and adaptive insulation optimise façade's performance and occupant comfort. Architects can maximise sunlight penetration into the building, reducing the reliance on artificial lighting by using light-redirecting devices. Incorporating sustainable materials into façade design is key to sustainability. Architects can assess the environmental impacts of façade materials and construction methods. Net-zero energy façades are the pinnacle of sustainable building design. These façades produce as much energy as they use, often through integrated renewable energy sources. Net-zero energy façades demonstrate a commitment to environmental responsibility and mitigating climate change. Climate-responsive façade design tailors buildings to local climatic conditions. Using weather patterns, architects can design façades that adapt to varying temperatures, sun exposure and wind [28].

4. Leveraging façade sensors for a responsive design

4.1 Designing adaptable spaces

4.1.1 Dynamic façades that respond to environmental conditions

Dynamic façades in architecture characterise an important development in the area, where building externals are designed to react smartly to dynamic weather conditions. These façades are designed to communicate with environmental factors such as sunlight, temperature and wind, demonstrating a state-of-the-art combination of technology, sustainability and aesthetics [4]. This architectural approach supports diverse purposes, ranging from improving energy efficiency to creating appealing visual experiences. The core idea behind dynamic façades is to design building envelopes that are capable of adapting themselves to enhance the indoor environment. Such flexibility can greatly influence a building's energy utilisation and thermal comfort, aligning with the overall objectives of sustainable architecture. The components used in dynamic façades differ, integrating several technologies and design methods. One strategy includes the application of photochromic and thermochromic materials in façade mechanisms. These materials respond to variations in luminous intensity and temperature, triggering the façade's appearance to change. Another method integrates kinetic components, such as movable louvres, panels and shading devices. These components can be manually regulated or automated to control parameters such as light infiltration, ventilation and sensible heat gain.

Researchers have shown a growing attention to dynamic façades. Bedon et al. [36] discusses the structural aspects and performance assessment of adaptive façades in modern buildings. Adaptive façades are designed to respond to changing conditions and improve the overall building performance. They highlight the need for experimental methods and regulations to evaluate the structural safety, durability and fire safety of these innovative façade systems. It also presents a classification proposal and possible metrics for assessing their structural performance. The chapter emphasises the importance of considering material-related, kinematic, geometrical and mechanical aspects in the design of adaptive façades. The goal is to develop standardised and reliable procedures for the mechanical and thermo-physical characterisation of these novel structural systems. In the same vein, Bedon et al. [47] discuss the importance of experimental testing for adaptive façades, which are building enclosures that can respond to changing conditions. It discusses the performance requirements of façades, including airtightness, water-permeability, fire resistance and structural performance. The study explains that testing can be done in a laboratory or on-site, and that the configuration of the testing should include all relevant details for performance assessment. It also discusses the challenges of testing adaptive façades, such as determining the limit deformations and addressing impact and blast load scenarios. The article concludes by emphasising the importance of certified facilities for testing adaptive façades. Sudhakaran et al. [48] studied the performance of a dynamic façade system in campus buildings. This research involved the application of a climate-adaptive building envelope on a base model and validates and analyses it through thermal simulation and prototype experimentation. The results indicated a noticeable projected energy saving of more than half in annual energy usage (approximately 60% less) as against the normal condition without the adaptive building envelope. It proves that the adoption of adaptive building envelopes that are tailored to the solar movement, incident solar radiation and summer and winter conditions show an improved functioning of the building envelope.

4.1.2 Creating flexible interiors for changing occupancy needs

Flexible interiors improve the functionality of a space and also contribute to sustainability, as they can prolong the life of a building and reduce the need for costly maintenance. One key aspect of creating flexible interiors is the incorporation of adaptable spatial configurations. This comprises the use of movable partitions, furniture and flexible partitioning approaches that allow spaces to be easily reconfigured to satisfy diverse needs. Architects have increasingly turned to solutions such as demountable walls, sliding components and convertible furniture to allow rapid alterations of spaces. These elements enable the swift adaptation of space, for example, an open office into individual workstations, a conference room or a lounge, based on specific requirements. This approach is in sync with the principles of sustainable architecture by reducing the need for new construction when occupancy requirements change. According to Zhang et al. [49] flexible interiors not only reduce construction waste but also lead to lower energy consumption and reduced greenhouse gases associated with construction activities, demonstrating that such design approaches contribute to the sustainability of buildings.

Additionally, the integration of smart building technologies plays a crucial role in achieving adaptable interiors. Sensors, automation and building management systems can be installed to monitor space utilisation and occupancy patterns in real time. This data can inform the dynamic changes in lighting, temperature and ventilation, creating an environment tailored to the existing occupants. Such intelligent systems boost user experience and support the prudent use of space. The benefits of flexible interiors extend beyond commercial buildings. In residential buildings, adaptable interiors allow landlords to reallocate spaces to accommodate changes in family sizes, lifestyle preferences or the need for remote workspaces. This adaptability increases the long-term value of residential properties and aids sustainable living practises.

4.2 Personalising user experience based on real-time data

Real-time data analytics is a vehicle for personalising user experiences. Sensors and Internet of Things (IoT) devices installed in buildings incessantly gather data on user behaviours, temperature, lighting and user preferences. This pool of data serves as the basis for designing spaces that adjust dynamically to the needs of each user. Real-time data permits the customisation of many features of the built environment. For example, smart lighting systems can alter colour intensity based on the time of day, occupancy in a room or user preferences. Likewise, heating, ventilation and air conditioning (HVAC) systems can be adjusted to maintain ideal thermal comfort for each occupant. This level of customisation improves user comfort and also contributes to energy efficiency by curtailing unnecessary energy consumption.

Furthermore, personalising the user experience is more important than environmental controls. Architectural designs are increasingly integrating responsive systems such as flexible partitions and spatial configurations that can be modified to suit particular applications. These systems can be controlled manually or automatically on the basis of data inputs. For instance, an office building might reconfigure itself into a conference room or workstation depending on the user's needs. Neuhofer et al. [50] identify the requirements of smart technologies for experience creation, including information aggregation, mobile connectedness and real-time synchronisation. It also highlights how smart technology integration can lead to two levels of personalised tourism experiences.

5. Shaping the future of intelligent architecture

5.1 Optimising building performance

5.1.1 Using sensor data to enhance building operations

Sensor devices have become a vital component in the pursuit of intelligent building practises. These devices can monitor a plethora of critical factors, such as temperature, humidity, air quality, lighting and electricity consumption. The data collected by these sensors serves as a valuable resource for architects, builders and facility managers to gain a deeper understanding of the dynamic of building use. One of the advantages of sensor data is that it enhances energy efficiency. For example, temperature and occupancy sensors can work together to regulate the heating, ventilation and air conditioning (HVAC) systems. When no building is occupied, this system can automatically reduce energy usage. Dong et al. [51] demonstrated substantial energy savings through sensor-driven HVAC optimisation. They found that buildings integrated with such systems gain a significant reduction in energy consumption, resulting in both cost savings and less carbon monoxide emissions.

Advanced lighting control systems, for instance, can alter brightness based on the natural circadian rhythms of occupants, thus improving comfort. Moreover, data on indoor air quality can activate ventilation systems to dissipate fresh air when contaminants pass threshold levels, contributing to a better indoor setting. Also, the addition of sensor extends to proactive maintenance. Sensors implanted in critical building systems, such as elevators, HVAC systems and electrical systems, endlessly monitor their performance. By evaluating this data, facility managers can predict when systems are expected to fail, allowing preventive maintenance and preventing downtime. The application of sensor data in architecture can also be scaled to create smarter and efficient urban cities. Cities are increasingly installing sensors to monitor traffic flow, energy usage, waste management and air quality. This data-driven approach informs urban planning and policymakers, paving the way for more sustainable cities.

5.1.2 Improving efficiency and reducing maintenance costs

Efficiency improvements in design often begin with the selection of façade materials. High-performance materials such as double-glazed windows, insulated panels and advanced coatings can significantly enhance the thermal performance of a building.

These materials aid in reducing heat loss during winter and heat gain during summer, thus improving energy efficiency. In addition, well-made façades can integrate passive solar systems, exploiting natural daylight while reducing glare and heat gain, consequently minimising the need for artificial lighting and cooling systems. Sheikh and Asghar [52] explore the design of an adaptive biomimetic façade for highly glazed buildings in hot and humid regions. The façade reduces solar heat gain and energy consumption while maintaining visual comfort. The design is inspired by the *Oxalis oregana* leaf, which tracks the sun's path and adjusts its position accordingly. The façade module can be folded horizontally and vertically, providing shading under both high and low sun angles. A case study of a 20-story office building in Lahore, Pakistan, demonstrates that retrofitting the façade reduces energy load by 32% and maintains recommended lighting levels in 50% of the interior space. The biomimetic façade offers significant energy savings while preserving visual comfort.

Likewise, the choice of façade materials can impact the cost of maintenance. Durable and maintenance-free materials can reduce the frequency and cost of maintenance. For example, the application of cladding can extend the useful life of façade, thereby minimising the need for regular painting. Façade designs that integrate rain screens and drainage systems can also impede dampness. Integrating smart technologies into façades can further improve productivity and lower the maintenance costs. Automated shading mechanisms, for instance, can regulate changing sunlight, thus reducing HVAC load. Self-cleaning coatings, which can be useful for façade surfaces, can reduce the build-up of dirt, thereby minimising maintenance needs.

5.2 Advancing sustainability

5.2.1 Designing eco-friendly buildings using data-driven approaches

Data-driven approaches include the gathering and analysing of data to inform the design of building façades. Sensors are deployed to collect data on temperature, solar radiation, humidity and wind patterns. This data is then used to make informed decisions throughout the lifespan of a building, from inception to demolition. Through data-driven analysis, designers can determine the effective approaches for increasing natural daylight while reducing heat gain or loss. For example, computational simulations can model the trajectory of the sun during the day and across seasons, enabling designers to position windows and shading devices to enhance daylighting and reduce the need for artificial lighting and cooling. Hosseini et al. [53] underscore the benefits of data-driven daylighting approaches in façades. They assess the concept of an interactive façade that can dynamically adjust to optimise daylight and enhance occupant comfort. The study develops a kinetic interactive façade that can transform based on dynamic daylight and occupant position, improving visual comfort. Daylight parametric simulations demonstrate the high performance of the kinetic interactive façades in improving visual comfort and controlling solar radiation. The results highlight the multifunctional aspects of the façade, which can prevent thermal discomfort and improve occupant health.

5.2.2 Reducing carbon footprints and energy waste

Buildings are responsible for a significant share of global greenhouse gas emissions. In the United States, they account for over 35% of the overall energy utilised and greenhouse gas emissions [54]. Façades, as the main barrier between the interior of a building and the building's exterior, can overcome these environmental impacts. Energy waste is a direct impact of unproductive building practises. The poorly designed façades can lead to excessive heating and cooling loads, resulting in higher energy utilisation. Architects contribute to the energy efficiency of buildings by minimising energy consumption through façade design. Reducing carbon greenhouse gases aligns with sustainability goals. Sustainable architecture aims to design buildings that harmonise with their surroundings, use resources prudently and have less negative environmental impact. Green façades are the cornerstone of this approach.

Façades should integrate advanced insulation systems and techniques to reduce heat gain/loss. Good insulation minimises the need for heating and cooling, thus minimising the use of energy. Selecting energy-efficient glazing materials with low U-values and better solar heat gain coefficients can considerably enhance the performance of façade. Integrating vents into façades enables natural ventilation, lowering dependence on HVAC installations and reducing energy usage. Using exterior shading components such as sunshades, louvres and brise-soleil can prevent extreme heat gain while allowing natural lighting and reducing the demand for artificial lighting and ventilation. Incorporating renewable energy sources, such as photovoltaic cells into façade design can generate clean energy. Sensor-driven can be used as an automatic control for lighting, heating and ventilation based on user conditions. Assessment of façade materials and the construction technique through life cycle assessments to ensure they have minimal environmental impact over their intended life. Bui et al. [4] propose a computational optimisation approach to improve the energy efficiency of buildings through the design of adaptive façades. Adaptive façades can adjust their thermal and visible transmittance according to changing climatic conditions. The approach combines a building energy simulation programme with an optimisation technique to design the adaptive façade system. The modified firefly algorithm is used in this study, but the method is not limited to a specific optimisation tool or building type. The proposed adaptive façade system is validated through two case studies, showing energy consumption reductions of 14.9–29.0% and 14.2–22.3% compared to static façades. This highlights the potential of adaptive façades to enhance building energy efficiency.

5.3 Innovating occupant-specific buildings

5.3.1 Designing spaces that are tailored to individual needs and preferences

The pursuit of designing spaces personalised to person's preferences denotes a fundamental shift in design philosophy. This is evident in the design of building façades, which serve as a link between the built environment and its occupants. Contemporary architects recognise that tailored spaces not only improve user satisfaction but also contribute to enhanced health and safety. Individual spaces are intrinsically user-specific, providing buildings that suit the unique requirements of users. This improved individual experience contributes to comfort, satisfaction and a sense of ownership over the space. Designing façades with personalisation in mind helps improve space allocation. Spaces can be modified for numerous functions, accommodating different tasks and dynamic requirements with minor refurbishment. Individualised spaces have been shown to positively impact welfare and productivity. For example, a well-lit space with adaptable lighting can minimise eye strain and increase awareness, whereas customisable interiors help relaxation. Spaces that are personalised to person's preferences tend to be used more efficiently. This can reduce resource utilisation of energy, water and materials are assigned based on actual needs rather than standardised norms [55].

User-specific design begin by conducting extensive user research to understand the specific requirements of the building's occupants. This information underpins the design decisions. Integrating spatial shapes that allow easy adaptation. Use partitions, modular furniture and versatile zoning strategies that can be adjusted to accommodate different activities and user preferences. Customisable façade elements that enable customisation, for example integrating windows, adjustable shading devices and balcony spaces that can be tailored to suit person's requirements. Material and finish choices offer options for interior elements such as the floor, wall and cabinet. This enables users to choose finishes that match their needs. Lee et al. [56] assessed control strategy for adaptive façades, specifically movable shading devices. The aim is to determine the optimal positions of the shades based on various control objectives such as daylighting, thermal comfort, glare prevention and energy conservation. Lee et al. [56] propose a multi-purpose control strategy that considers all these factors and aims to optimise heating, cooling, lighting energy and glare. The strategy aims to provide an effective and efficient solution for controlling adaptive façades.

5.3.2 Enhancing productivity and well-being

Increasing the use of daylight is a vital aspect of façade design. Well-placed windows, skylights and glass façades permit natural light to infiltrate into internal spaces, minimising the need for man-made lighting. Exposure to daylight has been associated with enhanced mood, reduced stress and improved productivity among building users. Façades that offer access to views of the natural environment, such as green areas, parks and water bodies, can have a positive impact on welfare. Research demonstrated that views of nature can reduce mental fatigue and improve cognitive function. Façades play a critical role in adjusting the thermal comfort of a building. Better insulation, shading materials and cooling techniques can help achieve thermal comfort annually. A comfortable indoor environment supports productivity and mitigates elevated temperature-based health problems. Façades can also contribute to acoustic performance by decreasing outdoor noise penetration. Noise-resistance components and glazing can provide a noiseless internal environment, minimise disturbance and anxiety and improve awareness and comfort. The aesthetic qualities of a façade, including its design, colour and texture, can influence persons' perception of the space. An appealing façade can offer a sense of self-importance and identity among individuals, positively impacting their welfare. Allowing occupants some level of control over ventilation, and lighting can contribute to a sense of comfort. Customised control allows occupants to regulate their environment to suit their needs, thus, improving well-being [57]. Hongisto et al. [58] investigate the relationship between the physical environment of an open-plan office and employee satisfaction. The researchers conducted a quasi-field experiment in a 135-employee office, where various refurbishments were made to improve thermal conditions, visual and acoustic privacy, ergonomics, interior design and spatial density. All employees were surveyed twice, before and after the refurbishment, and physical measurements were taken. The study sought to provide evidence of the impact of the office environment on job.

6. Conclusion and areas of future research

6.1 Conclusion

In conclusion, the incorporation of façade sensors into intelligent architecture represents a transformative leap forward in the way buildings are designed, constructed, operated and maintained. Façade sensors, including temperature, humidity, light, air quality, solar radiation and occupancy sensors, provide architects with a wealth of real-time data that empowers them to design buildings that are not only energy-efficient and sustainable but also user-centric and responsive to environmental conditions. The use of these sensors enables architects to make data-driven design decisions, resulting in more efficient and environmentally friendly buildings. By harnessing the capability of sensor data, architects can optimise energy consumption, improve indoor comfort and reduce their carbon footprint. These sensors facilitate the development of dynamic façades that adjust to vary weather conditions, offering occupants a more comfortable and visually aesthetic environment. In addition, the seamless integration of sensor data within architectural design process, fosters teamwork and interdisciplinary approaches, bringing together design, technology and sustainability. However, it presents challenges such as the control of a large volume of data and the need for architects to embrace computational design approaches and data visualisation techniques.

As we dive into the future of intelligent architecture, it is clear that façade sensors will continue to play a critical role in configuring buildings that are at the core of sustainability, user experience and productivity. By leveraging the insights gained from sensor data, architects can design buildings that satisfy current needs and contribute to a more sustainable and resilient future. The journey towards intelligent architecture is ongoing, and façade sensors will remain at the forefront of this transformative evolution, driving innovation and enhancing the built environment for generations to come.

6.2 Further research

There are numerous areas for further research that can offer a better understanding and extend the application of façade sensors in intelligent architecture. Further studies can investigate the following: [1] how nanoscale sensors can provide real-time data on structural health, air quality and other parameters without changing the appearance of the façade, [2] how 3D printing can be used to produce personalised building components with built-in sensors, allowing for highly personalised and responsive architectural designs and [3] how blockchain can improve data integrity, privacy and transparency in multistakeholder environments. Finally, how sensor-installed buildings can contribute to early warning systems, rapid response and post-disaster recovery efforts, thereby improving urban resilience.

Author details

Mubarak Reme Ibrahim
Baze University, Abuja, Nigeria

*Address all correspondence to: mubarak.ibrahim@bazeuniversity.edu.ng

References

[1] Heidari Matin N, Eydgahi A. Technologies used in responsive facade systems: A comparative study. Intelligent Buildings International. 2022;**14**(1):54-73

[2] Graamans L, Tenpierik M, van den Dobbelsteen A, Stanghellini C. Plant factories: Reducing energy demand at high internal heat loads through facade design. Applied Energy. 2020;**262**:114544

[3] Khan DS, Kolarik J, Hviid CA, Weitzmann P. Method for long-term mapping of occupancy patterns in open-plan and single office spaces by using passive-infrared (PIR) sensors mounted below desks. Energy and Buildings. 2021;**230**:110534

[4] Bui DK, Nguyen TN, Ghazlan A, Ngo NT, Ngo TD. Enhancing building energy efficiency by adaptive façade: A computational optimization approach. Applied Energy. 2020;**265**:114797

[5] Tian Z, Zhang X, Wei S, Du S, Shi X. A review of data-driven building performance analysis and design on big on-site building performance data. Journal of Building Engineering. 2021;**41**:102706

[6] Attia S, Lioure R, Declaude Q. Future trends and main concepts of adaptive facade systems. Energy Science & Engineering. 2020;**8**(9):3255-3272

[7] Fabbri K, Gaspari J, Bartoletti S, Antonini E. Effect of facade reflectance on outdoor microclimate: An Italian case study. Sustainable Cities and Society. 2020;**54**:101984

[8] Zheng X, Montazeri H, Blocken B. CFD analysis of the impact of geometrical characteristics of building balconies on near-façade wind flow and surface pressure. Building and Environment. 2021;**200**:107904

[9] Sawyer AO, Navvab M, Weissman D, Ji G. Facade photometry: Linking annual daylight performance to facade design. Buildings. 2022;**12**(10):1556

[10] Convertino F, Vox G, Schettini E. Evaluation of the cooling effect provided by a green façade as nature-based system for buildings. Building and Environment. 2021;**203**:108099

[11] Martinez A, Choi JH. Exploring the potential use of building facade information to estimate energy performance. Sustainable Cities and Society. 2017;**35**:511-521

[12] Mylonas A, Kazanci OB, Andersen RK, Olesen BW. Capabilities and limitations of wireless CO_2, temperature and relative humidity sensors. Building and Environment. 2019;**154**:362-374

[13] Attoue N, Shahrour I, Younes R. Smart building: Use of the artificial neural network approach for indoor temperature forecasting. Energies. 2018;**11**(2):395

[14] Mba L, Meukam P, Kemajou A. Application of artificial neural network for predicting hourly indoor air temperature and relative humidity in modern building in humid region. Energy and Buildings. 2016;**121**:32-42

[15] Woodstock TK, Karlicek RF. RGB color sensors for occupant detection: An alternative to PIR sensors. IEEE Sensors Journal. 2020;**20**(20):12364-12373

[16] Joyce DS, Zele AJ, Feigl B, Adhikari P. The accuracy of artificial

and natural light measurements by actigraphs. Journal of Sleep Research. 2020;**29**(5):e12963

[17] Montiel-Santiago FJ, Hermoso-Orzáez MJ, Terrados-Cepeda J. Sustainability and energy efficiency: BIM 6D. Study of the BIM methodology applied to hospital buildings. Value of interior lighting and daylight in energy simulation. Sustainability. 2020;**12**(14):5731

[18] Boyce P, Hunter C, Howlett O. The Benefits of Daylight through Windows. Troy NY: Rensselaer Polytech Inst; 2003

[19] Shirazi A, Newton S, Christensen PH. Processing and configuring smart and sustainable building management practices in a University Building in Australia. Sustainability. 2023;**15**(7):6194

[20] Canale L, Di Fazio AR, Russo M, Frattolillo A, Dell'Isola M. An overview on functional integration of hybrid renewable energy systems in multi-energy buildings. Energies. 2021;**14**(4):1078

[21] Attia S, Favoino F, Loonen R, Petrovski A, Monge-Barrio A. Adaptive façades system assessment: An initial review. In: 10th Conference on Advanced Building Skins. Bern, Switserland: Economic Forum; 3-4 Nov 2015. pp. 1275-1283

[22] O'Grady T, Chong HY, Morrison GM. A systematic review and meta-analysis of building automation systems. Building and Environment. 2021;**195**:107770

[23] Demanega I, Mujan I, Singer BC, Anđelković AS, Babich F, Licina D. Performance assessment of low-cost environmental monitors and single sensors under variable indoor air quality and thermal conditions. Building and Environment. 2021;**187**:107415

[24] Gonzalez-Martin J, Kraakman NJR, Perez C, Lebrero R, Munoz R. A state–of–the-art review on indoor air pollution and strategies for indoor air pollution control. Chemosphere. 2021;**262**:128376

[25] Nugraha M, Adriansyah A. Optimization of sensor model for solar radiation measurement with a pyranometer. IOP Publishing. 2021;**739**:012080

[26] Qiu Z, Wang J, Yu B, Liao L, Li J. Identification of passive solar design determinants in office building envelopes in hot and humid climates using data mining techniques. Building and Environment. 2021;**196**:107566

[27] Vassiliades C, Michael A, Savvides A, Kalogirou S. Improvement of passive behaviour of existing buildings through the integration of active solar energy systems. Energy. 2018;**163**:1178-1192

[28] Sun K, Zhao Q, Zhang Z, Hu X. Indoor occupancy measurement by the fusion of motion detection and static estimation. Energy and Buildings. 2022;**254**:111593

[29] Dong B, Andrews B, Lam KP, Höynck M, Zhang R, Chiou YS, et al. An information technology enabled sustainability test-bed (ITEST) for occupancy detection through an environmental sensing network. Energy Build. 2010;**42**(7):1038-1046

[30] Ziaja D, Nazarko P. SHM System for Anomaly Detection of Bolted Joints in Engineering Structures. Amsterdam: Elsevier; 2021. pp. 3877-3884

[31] Bedon C, Santi MV, Fasan M. Considerations on efficient procedural steps for seismic capacity assessment and diagnostics of historic structural glass systems. Soil Dynamics and Earthquake Engineering. 2022;**163**:107562

[32] Bedon C, Noè S, Fasan M, Amadio C. Role of in-field experimental diagnostic analysis for the derivation of residual capacity indexes in existing pedestrian glass systems. Buildings. 2023;**13**(3):754

[33] Han K, Zhang J. Energy-saving building system integration with a smart and low-cost sensing/control network for sustainable and healthy living environments: Demonstration case study. Energy and Buildings. 2020;**214**:109861

[34] Pillai DS, Shabunko V, Krishna A. A comprehensive review on building integrated photovoltaic systems: Emphasis to technological advancements, outdoor testing, and predictive maintenance. Renewable and Sustainable Energy Reviews. 2022;**156**:111946

[35] Borile S, Pandharipande A, Caicedo D, Schenato L, Cenedese A. A data-driven daylight estimation approach to lighting control. IEEE Access. 2017;**5**:21461-21471

[36] Bedon C, Honfi D, Machalická KV, Eliášová M, Vokáč M, Kozłowski M, et al. Structural characterisation of adaptive facades in Europe–part I: Insight on classification rules, performance metrics and design methods. Journal of Building Engineering. 2019;**25**:100721

[37] Santamouris M, Ding L, Osmond P. Urban heat island mitigation. In: Decarbonising the Built Environment: Charting the Transition. Springer; 2019. pp. 337-355

[38] Bedon C, Zhang X, Santos F, Honfi D, Kozłowski M, Arrigoni M, et al. Performance of structural glass facades under extreme loads–design methods, existing research, current issues and trends. Construction and Building Materials. 2018;**163**:921-937

[39] Peng C, Feng D, Guo S. Material selection in green design: A method combining DEA and TOPSIS. Sustainability. 2021;**13**(10):5497

[40] Yang X, Peng LL, Jiang Z, Chen Y, Yao L, He Y, et al. Impact of urban heat island on energy demand in buildings: Local climate zones in Nanjing. Applied Energy. 2020;**260**:114279

[41] Farah S, Whaley D, Saman W, Boland J. Integrating climate change into meteorological weather data for building energy simulation. Energy and Buildings. 2019;**183**:749-760

[42] Pedersen TH, Nielsen KU, Petersen S. Method for room occupancy detection based on trajectory of indoor climate sensor data. Building and Environment. 2017;**115**:147-156

[43] Kim H, Ahn H. Temporary facility planning of a construction project using BIM (building information modeling). Computing in Civil Engineering. 2011;**2011**:627-634

[44] Niu S, Pan W, Zhao Y. A virtual reality integrated design approach to improving occupancy information integrity for closing the building energy performance gap. Sustainable Cities and Society. 2016;**27**:275-286

[45] Yang E, Bayapu I. Big data analytics and facilities management: A case study. Facilities. 2020;**38**(3/4):268-281

[46] Liu C, Zhang G, Arıcı M, Bian J, Li D. Thermal performance of non-ventilated multilayer glazing facades filled with phase change material. Solar Energy. 2019;**177**:464-470

[47] Bedon C, Honfi D, Machalická KV, Eliášová M, Vokáč M, Kozłowski M, et al. Structural characterisation of adaptive facades in Europe-part II: Validity of

conventional experimental testing methods and key issues. Journal of Building Engineering. 2019;**25**:100797

[48] Sudhakaran P, Joshi B, Singh G, Kumar M, Ansari S. Performance simulation and validation of climate adaptive dynamic facades in campus buildings. International Journal of Advanced Science and Technology. 2020;**29**(6):4721-4736

[49] Zhang Y, Yan D, Hu S, Guo S. Modelling of energy consumption and carbon emission from the building construction sector in China, a process-based LCA approach. Energy Policy. 2019;**134**:110949

[50] Neuhofer B, Buhalis D, Ladkin A. Smart technologies for personalized experiences: A case study in the hospitality domain. Electronic Markets. 2015;**25**:243-254

[51] Dong J, Winstead C, Nutaro J, Kuruganti T. Occupancy-based HVAC control with short-term occupancy prediction algorithms for energy-efficient buildings. Energies. 2018;**11**(9):2427

[52] Sheikh WT, Asghar Q. Adaptive biomimetic facades: Enhancing energy efficiency of highly glazed buildings. Frontiers of Architectural Research. 2019;**8**(3):319-331

[53] Hosseini SM, Mohammadi M, Guerra-Santin O. Interactive kinetic façade: Improving visual comfort based on dynamic daylight and occupant's positions by 2D and 3D shape changes. Building and Environment. 2019;**165**:106396

[54] Kampelis N, Papayiannis GI, Kolokotsa D, Galanis GN, Isidori D, Cristalli C, et al. An integrated energy simulation model for buildings. Energies. 2020;**13**(5):1170

[55] Sakellaris IA, Saraga DE, Mandin C, Roda C, Fossati S, De Kluizenaar Y, et al. Perceived indoor environment and occupants' comfort in European "modern" office buildings: The OFFICAIR study. International Journal of Environmental Research and Public Health. 2016;**13**(5):444

[56] Lee D, Cho YH, Jo JH. Assessment of control strategy of adaptive façades for heating, cooling, lighting energy conservation and glare prevention. Energy and Buildings. 2021;**235**:110739

[57] Khademagha P, Aries M, Rosemann A, Van Loenen E. Implementing non-image-forming effects of light in the built environment: A review on what we need. Building and Environment. 2016;**108**:263-272

[58] Hongisto V, Haapakangas A, Varjo J, Helenius R, Koskela H. Refurbishment of an open-plan office–environmental and job satisfaction. Journal of Environmental Psychology. 2016;**45**:176-191

Chapter 5

Adaptive Textile Façade Systems – The Experimental Works at D1244

Lucio Blandini, Christina Eisenbarth, Walter Haase, Moon-Young Jeong, Michael Voigt, Daniel Roth, Arina Cazan and Maria Matheou

Abstract

Adaptive façade systems are a promising approach to achieve a dynamic response to varying weather conditions and to individual user demands. Within the framework of the Collaborative Research Center (CRC) 1244 at the University of Stuttgart the use of adaptive systems and the related architectural potential is explored with the aim of reducing the consumption of natural resources as well as waste generation and hazardous emissions. The targeted parameters for the façade design include solar radiation, temperature, wind speed, relative humidity, daylighting, and user interaction. To generate an experimental platform for the research work, a 36.5 m high adaptive experimental tower, D1244, has been designed and built on the University campus. The temporary façade of the tower is currently being replaced floor by floor, in order to validate different research approaches. The first implemented façades focus on textile systems, because of their lightweight and the different functions that can be easily integrated. Further material systems will be investigated in the next future.

Keywords: adaptivity, textile solutions, resilience, interaction, kinetic architecture

1. Introduction

The Collaborative Research Center (CRC) 1244 "Adaptive Skins and Structures for the Built Environment of Tomorrow" at the University of Stuttgart has been working since 2017 on the question of how future built environments can be created reducing the use of resources and the associated greenhouse gas emissions (GHG). The building sector is responsible for more than 50% of global resource consumption and for more than 38% of global CO_2 emissions [1]. The target of the interdisciplinary program is the development of new design strategies and technologies, which enable structures and envelopes to be adaptive against loading and environmental actions. The research group comprises architects as well as structural, mechanical, control, and aeronautical engineers and computer scientists.

Within the scope of CRC 1244, adaptive structures and façades are understood as systems whose physical properties are actively manipulated by means of control systems. The state of the system is monitored by sensors. In the specific case of façades, the overall target is the design and validation of systems that enable the manipulation of transparency, reflectivity, humidity rate, insulation, cooling and acoustic properties, in order to control indoor as well as outdoor conditions in the vicinity of the building envelope.

Conventional envelopes can only provide a very limited range of reactions to varying external agents or to changing user needs. When using such systems, façade engineers often can achieve sub-optimal design [2]. While other researchers have been in the past investigating performances and design methods of adaptive skins in general [3], in this chapter the focus is on a specific field: adaptive textile façades. The reason for this focus is the lightweight character of such façade types and the potential of integrating and combining different functions while keeping a low ecological footprint.

CRC 1244 has set a strong experimental background for the research work. A 36.5 m high adaptive experimental tower, called D1244, has been built to test the proposed approaches on a large-scale experimental structure that offers real-world conditions (**Figure 1**). D1244 is the world's first adaptive high-rise building [4]. The unique feature of this demonstrator is the integration of sensors and adaptive components into the load-bearing structure and skin. All elements are assembled in such a way that they can be later substituted without generating any waste. The adaptive components in the load-bearing structure enable it to react autonomously against external disturbances such as winds and earthquakes. The building envelope currently consists of a single-layer recycled membrane that is being gradually replaced by adaptive façades as the research projects unfold.

This chapter focuses on three systems: *HydroSKIN* on the tenth floor, addressing rainwater collecting strategies and the potential of evaporative cooling; *FiberSKIN* and *MagneticSKIN* on the ground floor which are targeting the interaction between

Figure 1.
View of the experimental high-rise building D1244, Stuttgart © R. Müller.

people with inside and outside spaces. Additionally, it provides an overview of further textile systems such as *KineticSKIN*, which is currently in the design stage.

2. HydroSKIN

HydroSKIN represents the first lightweight building skin, that collects the wind-driven rainwater hitting the building façade and releases water in heat periods to cool the interior and exterior environment by evaporation (**Figure 2**). The aim is a drastic reduction of urban inundation and heat risks by relieving the sewage infrastructure as well as providing natural microclimate regulation with a minimal amount of embedded mass, energy, and emissions.

2.1 Climate context

Heavy rainfall events and extreme heat are becoming more intense, frequent, and long-lasting [5]. The increasing urban densification with coherent surface sealing in urban agglomerations enhances precipitation runoff on the one hand, as well as solar radiation absorption, thus causing urban heat island effects on the other hand. While social developments lead to increasing urban densification, surface sealing, and the construction of high-rise buildings, the effects of climate change, such as extreme heat and heavy rainfall, require the opposite: the creation of more permeable surfaces and buffer areas for reducing inundation and heat exposure. The average annual ratio between evaporation and runoff for non-built-up surfaces, such as green areas,

Figure 2.
HydroSKIN prototypes at the D1244 experimental high-rise building in Stuttgart © S. Cichowicz.

is about 60% evapotranspiration, 25% groundwater recharge, and 15% rainwater runoff. In comparison, sealed surfaces demonstrate an average runoff of over 90% [6]. In conclusion, the aim is to approximate the water balance of built-up areas to that of non-built-up areas by reducing the precipitation runoff after heavy rainfall events, as well as by increasing evaporation and latent cooling in urban areas.

To avoid irreversible damage to humans and the environment, the Intergovernmental Panel on Climate Change is pursuing two complementary approaches, based on the concepts of mitigation and adaption. The first aim is to reduce climate change by significant and sustained reductions in greenhouse gasses (mitigation). However, since climatic consequences are to be expected even with zero manmade CO_2 emissions, strategies and technologies for adaptation to the expected climate situation are being developed [5].

The combination of climate mitigation and adaptation strategies, by addressing both climate challenges of urban heat islands as well as pluvial inundation risks, is seen to have great potential for dealing with global environmental issues sustainably and effectively.

2.2 Concept

Most façade systems only focus on the building-physical performance. *HydroSKIN* faces the current climatic challenges by addressing both climate mitigation and adaptation strategies with a new type of functionalities in the building skin (**Figure 3**). The incorporation of textile and foil-based materials into the building skin opens a revolutionary new spectrum of performances in the façade: with a minimal weight per unit area, the use of a special, functionalized, multi-layered textile enables integration of decentralized rainwater harvesting into the façade, with time-delayed evaporative cooling of the building and its environment. The aim of the *HydroSKIN* is to improve drastically and sustainably the climate resilience of buildings and cities by simultaneously reducing precipitation runoff and inundation risks as well as urban heat island effects; achieving this with a minimum amount of technical effort, resource and energy consumption [7].

During heavy rainfall events accompanied by wind, the *HydroSKIN* add-on element absorbs the wind-driven rain striking the building façade. Thereby the

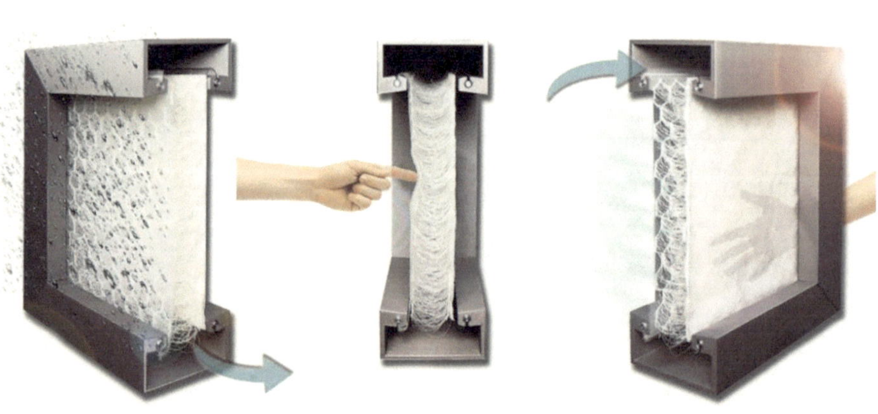

Figure 3.
HydroSKIN concept for rainwater harvesting of wind-driven rain and for evaporative cooling © C. Eisenbarth/ ILEK.

minimal material façade element reduces the load on urban sewage infrastructure and decreases the risk of flooding. Wind causes rain to have a horizontal velocity component and therefore increases the water impact on the façade. During hot periods, the absorbed rainwater can be targeted and time-delayed and released by wetting and evaporating on the façade. This improves the urban microclimate by cooling the building envelope and causing a down-flow of cold air into the urban area around the building thereby mitigating urban heat islands. The hybrid component addresses both rising climatic impacts of heat and inundation risks on urban architecture in a single hybrid envelopment solution. Depending on the prevailing climatic conditions, *HydroSKIN* can also be configured to perform only as one mono-functional device for rainwater harvesting or evaporative cooling [7–9].

2.3 System design

HydroSKIN is designed as a multi-layered textile structure to fulfill the multiple requirements of water absorption, storage, transport as well as evaporation (see **Figure 4**). The first layer provides a water-permeable filter facing to the outside. It protects the structure behind it from accumulation of dirt particles, insects, etc., thus favoring the water permeation into the multi-layered system by splitting the incoming raindrops. The second layer consists of a three-dimensional water-transporting spacer fabric whose pile threads on the one hand transport the incoming and outgoing water droplets and on the other hand favor air circulation by an open porous structure with large surface area, thereby enhancing the evaporative cooling performance.

An intermediate layer with high water absorbency can optionally be integrated to increase water storage capacity and evaporation duration of the textile multi-layer system in very hot and dry regions. The water-bearing layer, consisting e.g., of a foil, is on the inside and serves to provide water drainage and collection into the lower profile system. The individual layers are assembled by a force fit and are fixed in a frame profile system by means of textile joining techniques. The polymer-based textiles can be manufactured out of recycled material. Besides the textile mono-material system

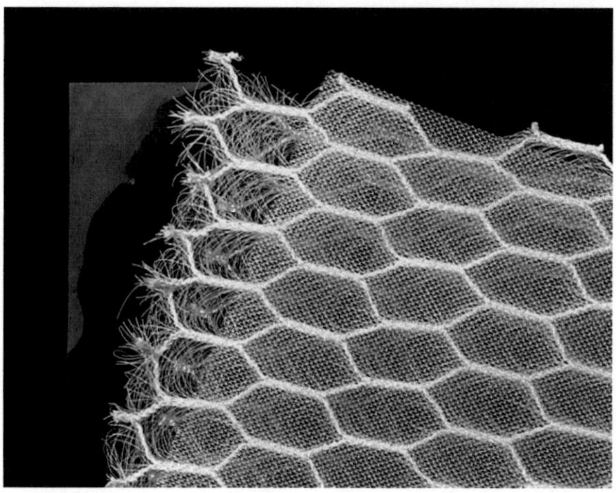

Figure 4.
Multi-layer system design of HydroSKIN © C. Eisenbarth/ILEK.

can be easily detached from the enclosing frame profile to return all system components to the material cycle. The water supply and discharge conduits are connected to the frame profile, enabling both the wetting of the *HydroSKIN* during hot periods and water drainage of the absorbed precipitation yields [9].

2.4 Potential

The advantages of façade-integrated rainwater harvesting consist not only in relieving the load on urban sewage infrastructure but also in reducing global freshwater consumption of residential buildings by up to 46% as well as in saving energy by up to 26% [7, 10].

Compared to conventional hard building surfaces, such as glass, optical investigations of the droplet impact behavior indicate a high permeability of textile materials. Evaporative cooling is one of the oldest and simplest principles of air-conditioning technology: the phase transition of water from the liquid to the gaseous state at temperatures below the boiling point extracts heat energy from the surrounding air. Frescoes from around 2500 BC show the fanning out of ceramic vessels filled with water, whose large pore content allows a large amount of water to be absorbed and evaporated on its surface [11, 12]. By the specifically adjustable surface structure and porosity of textiles, one can obtain a maximum surface area for water evaporation with a minimum amount of material. Thus, textiles are of particular interest for their application as evaporative cooling materials. The development of synthetic fibers since 1945 has even increased their evaporative cooling potential, since their large surface structure can be functionalized precisely [9, 13].

The evaporative cooling potential of water-saturated textile fabrics shown in **Figure 5** was investigated by empirical test series on an evaporation test bench under laboratory conditions of approx. 35°C room air temperature and 20–30% room air humidity. Humidification of the textile results in an immediate temperature reduction of 8–12 K, which under real weather conditions including wind velocity increases to

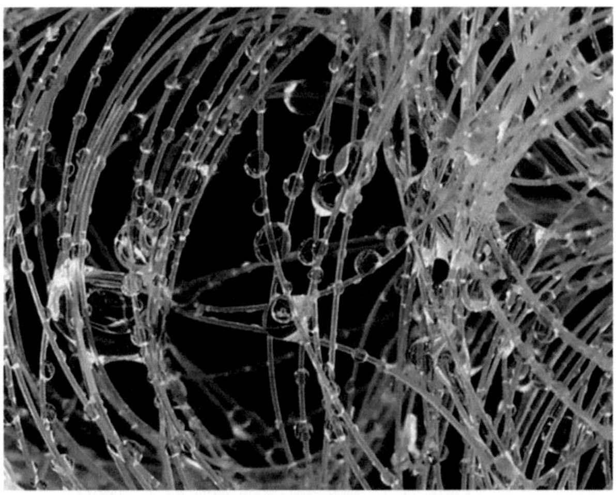

Figure 5.
Water droplets accumulating in three-dimensional spacer fabric © C. Eisenbarth/ILEK.

a façade surface temperature reduction of more than 20 K. The temperature decrease is accompanied by a coherent cool downdraft of about 0.2–0.4 m/s, which indicates a potentially useful implementation in tall buildings providing a cooling tower for the urban space below.

2.5 Implementation

HydroSKIN elements are designed to be applied to both new and existing buildings. Retrofitting of existing façades is limited to the static restriction of reducing the additive loads to a minimum, as these were not considered in the original design of the existing construction. The lightweight elements comply with such a constraint, still allowing for a wide range of design options and performance scenarios. The multi-layer design of *HydroSKIN* can be individually customized to the respective climate conditions, user requirements, and design guidelines. Creating a unique translucent esthetic effect with a textile, tactile surface texture, these climate-adaptive textile façades can be applied in a huge variety of design options, such as printed, colored, illuminated, 3D-shaped, or kinematic elements, that provide different states of shade and visibility in the façade from opaque closing over partial shading to fully transparency. The multi-layered collector and evaporator element provides a minimal weight per unit area of only approx. 1 kg/m^2 in a dry state and approx. 5 kg/m^2 in a water-saturated state, allowing it to be retrofitted to most conventional façade systems.

Further development of the *HydroSKIN* leads to a completely textile and film-based, multi-functional façade system [14]. In combination with an automated control and regulation strategy, the use of adaptive *HydroSKIN* façades can significantly improve indoor conditions and user comfort while reducing the consumption of water, materials, and energy at the same time [15].

Conventional high-rise buildings are characterized by a significant consumption of material and energy as well as high emission values, offering at best only marginal qualities for urban climate resilience. As Fazlur Khan once pointed out by his expression "Premium for height," the material consumption, as well as the embedded amount of "gray" energy bound up in the building, increases disproportionally with the building height due to the rising wind loads acting on the building façade [16]. On the other side, benefits result from the implementation of the *HydroSKIN* façades on high-rise buildings, as the first prototype façades at the D1244 building in Stuttgart demonstrate.

The façade surface of tall buildings such as skyscrapers offers not only a much larger absorption surface than its horizontal roof or ground surface. Simulations show, that above a building height of approx. 30 m, the amount of wind-driven rain yields per square meter façade surface is even greater than the amount of vertically falling precipitation per square meter on horizontal roof or ground surfaces. With the building height, wind speed rises, thus causing a stronger horizontal deflection of the precipitation drops and increasing wind-driven rain yields hitting the building façade [7]. During hot days, such wind velocities cause higher evaporation of water and enhance the cooling performance. Considering this potential of vertical retention and evaporation surfaces as a new "benefit for height" we wish for a new era of climate-adaptive and climate-resilient high-rise buildings (**Figure 6**).

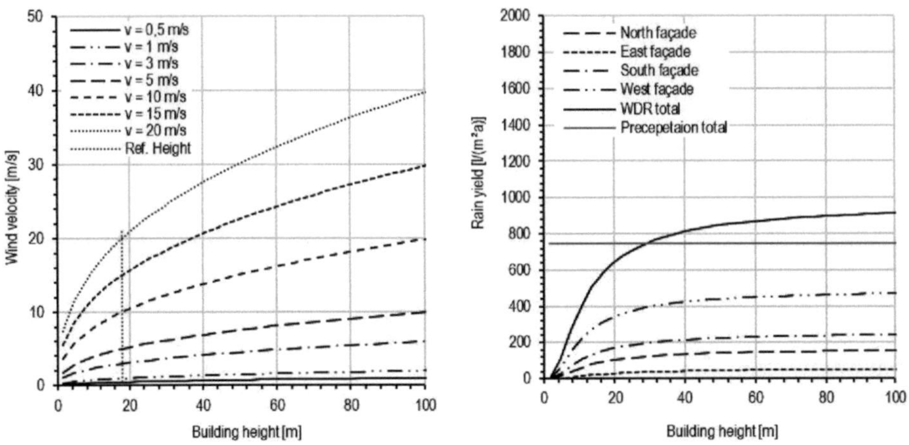

Figure 6.
Increase of wind speed (left) and wind-driven rainwater yields per square meter in comparison to vertically falling precipitation per square meter (right) with the building height [7].

3. FiberSKIN

3.1 Concept

The idea behind the design of *FiberSKIN* is a moving screen on the ground floor that allows for a smooth transition from the interior space, where the hydraulic structural system of D1244 is showcased, to the surrounding platform area, where events and various activities take place. An interdisciplinary team of architects, structural engineers, and mechanical engineers came up with a customized veil-like screen made of lightweight and fully recyclable glass and basalt fibers [17]. The project was also a methodological test to validate early interdisciplinary collaborations in the field of adaptive façades [18]. The tight link to the panel manufacturers allowed for a highly customized solution, especially in the definition of the fiber pattern. The developed solution has a wide range of potential applications where no thermal insulation is required and lightweight and movable panels are required.

Featuring the integration of lightweight textiles and a double-sliding mechanism, *FiberSKIN* blends the conventional notions of curtain wall and curtain (see **Figures** 7 and **8**): it prevents water penetration into the interior and regulates light transmission while generating special visual effects. The fiber panels cover three sides of the ground floor: two fixed panels clad on the southwest and northeast sides and two layers of movable panels clad on the southeast side. During the opening sequence, the permeability of the screen varies, creating an extraordinary spatial experience as the overlapping patterns of the two sliding panels constantly change. When closed, the panels overlap to provide a semitransparent screen; when fully open, the interior space is visually linked to the outdoors.

3.2 Panel design

The design is based on a geometric pattern radiating from intentionally placed clamping points. The number of clamping points differs depending on the function of the panels: the two 5.2 m wide movable panels are clamped with 17 nodes along

Figure 7.
View of FiberSKIN at the south corner of D1244 © M. Jeong/ILEK.

Figure 8.
View through the SW fixed panel from inside D1244 © M. Jeong/ILEK.

the horizontal edge to glide smoothly along the curved corners, while the 6.6 m long fixed panels only need 10 nodes to withstand wind forces (see **Figure 9**). In order to achieve a consistent pattern across all sides, a geometric rule is applied. This rule

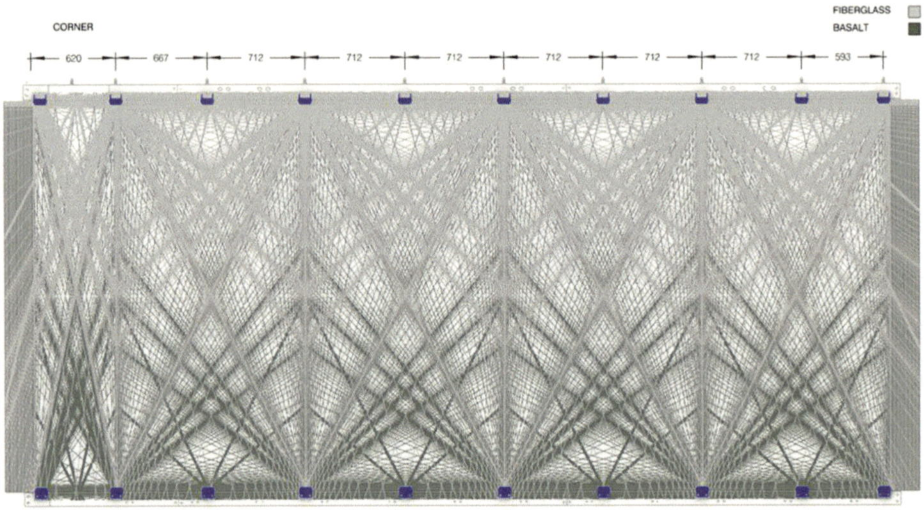

Figure 9.
Drawing for fixed panels and clamping pockets © M. Jeong/ILEK.

affects the pattern parametrically in response to variations in distance between nodes. The resulting pattern design leads to an extraordinary spatial experience indoors and outdoors based on the way the shadows are cast and the angle at which they are cast.

A color scheme inspired by the constituent materials of the fibers (basalt and glass) has been laid out to enhance the metaphorical level of the design. In the fixed panel both materials are laid together by engaging the i-Mesh fiber placement technology: the basalt fibers are concentrated in the bottom part and generate a metaphorical and optical link to the ground and the earth, given their volcanic origin. Meanwhile, glass fibers placed primarily at the top of the panel transmit considerably more light and are used to create a link to the sky. In the moving panel, the two material constituents are made visible in a different way since the two panels overlap in the closed position: the front panel is made entirely out of glass fibers and the back panel is made entirely out of basalt fibers.

3.3 Kinetic concept and mechanical implementation

The task for the kinetic concept was to arrange a semitransparent façade with a weather protection function, to be flexible enough to allow the interior to be fully opened for events. To meet these requirements, various methods were applied in an interdisciplinary manner between architects and mechanical engineers to find a variety of innovative solutions. With the help of brainwriting and the gallery method [19], more than 20 different opening concepts were developed within a very short period of time, which on the one hand were reminiscent of familiar openings such as theater curtains, but on the other hand also exhibited quite complex and organic movement patterns.

The concepts generated were then evaluated and selected based on various factors such as visual appearance and ease of implementation. The result was a concept in which two layers of textile move in front of each other, and the irregular distribution of fibers creates a superimposed interference effect. This also plays with the visibility of the building's interior technology, as some sections are more transparent than others, as well as with the incident light.

To implement the mechanical structure, some reference applications were studied at the beginning. For example, sailboats, garage doors, or conveyor systems in mechanical engineering have similar properties to those required for this façade. In order to gain an insight into the design and construction as well as the special features, manufacturers of each of these products were contacted and expert opinions on the transferability of this façade system were obtained. In the course of this exchange, industrial partners were also acquired to support the implementation of the adaptive façade with knowledge and technical components.

In the end, the mechanical system was inspired by the side sectional garage doors from Hörmann KG. These move in a similar way on the horizontal plane and their guide system was therefore a good reference. In addition, the thematic proximity to the construction industry was another advantage. The CAD model for the substructure of FiberSKIN is depicted in **Figure 10**. One major challenge in designing the façade was the transition from tolerances between the structure of the building (some cm) to the façade (a few mm). In order to achieve this, two specially designed support structures were integrated into the design. The first one is shown in **Figure 10** in the right upper corner. The sheet-metal design was selected as it meets the requirements for lightweight design and a high degree of design freedom. Here, another industry partner (TRUMPF Werkzeugmaschinen SE + Co. KG) was supporting to proper design of the brackets.

On the lower side of the façade, a classic L-profile was used, in which sufficient adjustment possibilities were provided by means of elongated holes in order to meet the small tolerances of the adaptive façade. As a measure to compensate for further tolerances and, in particular, to pre-tension the textile, a variety of roller carriers with corresponding compression springs were installed (see **Figure 10**, middle detail picture). These springs allow continuous adjustment of the pre-tensioning of the façade and are at the same time reliable even under high wind loads on the textile since they cannot overstretch in contrast to tension springs.

These combined measures jointly made it possible to meet a tolerance of approx. 2 mm over the entire width of 6 meters. Corresponding laboratory tests confirmed the design through endurance tests in which the façade underwent over 20,000 cycles.

The kinetic movement of the adaptive façade can be seen in the **Video 1**, https://bit.ly/46n1DgX.

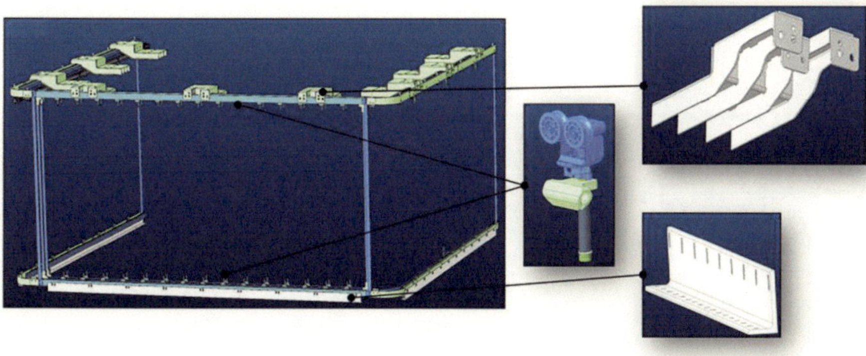

Figure 10.
CAD model of the mechanical substructure of FiberSKIN © M. Voigt/IKTD.

3.4 Details embedded in manufacture

In order to keep the engineering effort and the costs for the prototype as low as possible, while maintaining a high degree of design freedom, suitable reference applications were selected. The advantage here is that a large number of components have already been designed for similar use cases, which can only be slightly adapted and transferred for our prototype. A high degree of design freedom was otherwise allowed by the customized pattern design or by bespoke detailing. The focus was set on the clamps and on the large brackets that hold the façade. The clamps, which were placed regularly along the façade, served as an interface between the mechanical and architectural components. These are an integral part of the kinetic mechanism and fix the textile through special keder connections (see nodes blue marked in **Figure 9**).

The interdisciplinary and integrated design process of the clamps can be visualized based on their development. As these form the interface between the architectural part and the mechanical engineering part, they were subject to the most iterations. **Figure 11** visualizes the different development stages. It can be seen, that the design process started with a very rudimentary functional oriented CAD model of the assembly (**Figure 11**, Gen. (1). After the first discussions the improvements in the clamp design were mostly linked to lightweight design but also included some first shape finding aspects. Afterward, the shape was significantly improved in the third generation. Further improvements integrating new functionalities (Gen. 4 and 5, further connection possibilities were added) led finally to the stage where the clamps were integrated in the whole assembly back again. Generation 6 shows the detailed and final manufacturing version of the assembly including the clamps. This assembly was then used to frame and pre-tension the mesh. Therefore, on each of the blue-marked connection points in **Figure 9**, one of the clamp assemblies was mounted. Together with the mesh, this forms the two movable panels of the adaptive FiberSKIN façade.

4. MagneticSKIN

A user-centric approach in façade design involves placing the needs, preferences, and experiences of building occupants and users at the forefront of the design process. It emphasizes creating façades that not only fulfill functional requirements

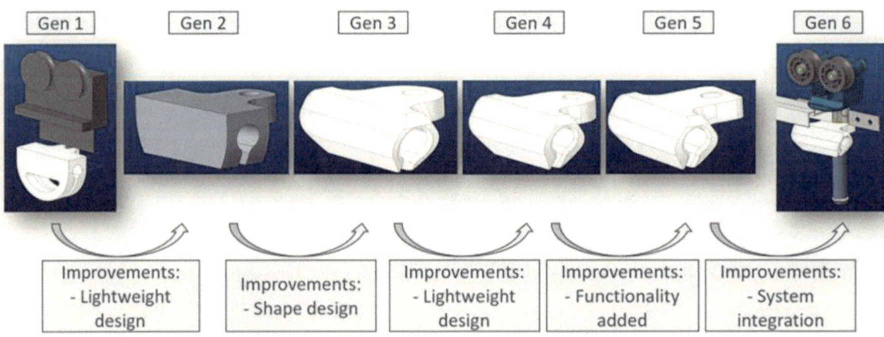

Figure 11.
Design process of the keder clamp–optimizing shape, weight, and functionality © M. Voigt/IKTD.

but also enhance the well-being, comfort, and overall satisfaction of the people who interact with the building skin.

4.1 Concept

MagneticSKIN is a pioneering interactive façade system designed to respond dynamically to human touch (see **Figure 12**). By exploring non-verbal ways of communication through haptic interaction, this approach aims to create a harmonious interplay between architectural esthetics and human experience. The ground level of D1244 is ideal for this purpose since it offers direct and convenient accessibility for individuals to engage with both the external and internal layers of the system.

In contemporary architectural practices, there is a growing emphasis on the ability to tailor specific properties of building envelopes to enhance comfort and overall space usability [20]. This focus primarily revolves around meeting physiological needs and individual preferences. However, the exploration of psychological needs and the broader activation of human senses remains largely uncharted territory. The proposed system delves into the significance of bridging this gap and investigates the potential benefits of integrating sensory stimuli to create more engaging and immersive built environments.

"Touch is the sensory mode that integrates our experience of the world with that of ourselves" according to Juhani Pallasmaa [21]. By considering touch and other sensory experiences during the design process, architects can enhance the emotional connection people have with their surroundings, leading to more memorable and enjoyable spaces [22]. The overall aim is to create a system that places the user at the core of the design process and to start understanding how haptic experiences influence perception, emotions, and behavior.

The interaction system follows the principles of system dynamics comprising of a set of sensors and actuators interconnected through a microcontroller and a specific

Figure 12.
Live interaction with the MagneticSKIN façade during the CRC 1244 symposium in May 2023 © U. Regenscheit.

set of rules or code. This design allows the system's behavior to be dynamically shaped by continuous interaction with users.

4.2 Pattern design

The arrangement of the round permanent magnets on the membrane surface follows a deliberate pattern inspired by the key trigger points found in an average-sized human hand. The abstract representation of these points results in a group of eight magnets. There is a total of five variations of this group, out of which the overall semi-regular clustered pattern is created by organically arranging them across the canvas (see **Figure 13**).

Each group of eight magnets corresponds to an electromagnet and sensor, working together to form what is referred to as an "active module." In addition to the active modules, there are "passive modules" composed of either individual permanent magnets or groups of magnets, to which no actuator is assigned. The role of these passive modules is to harmoniously integrate the pattern, especially in areas not easily reachable by hand for users interacting with the façade.

4.3 Detailing

The structural system consists of a wooden frame placed on adjustable steel posts for water protection and supported at the top by steel brackets (see **Figure 10**). By employing exclusively bolt/screw connections, the entire system can be easily disassembled, making it highly reusable and recyclable. The same principle applies to both inner and outer lightweight textile layers, which cover up the substructure, as well as to all components of the interaction system.

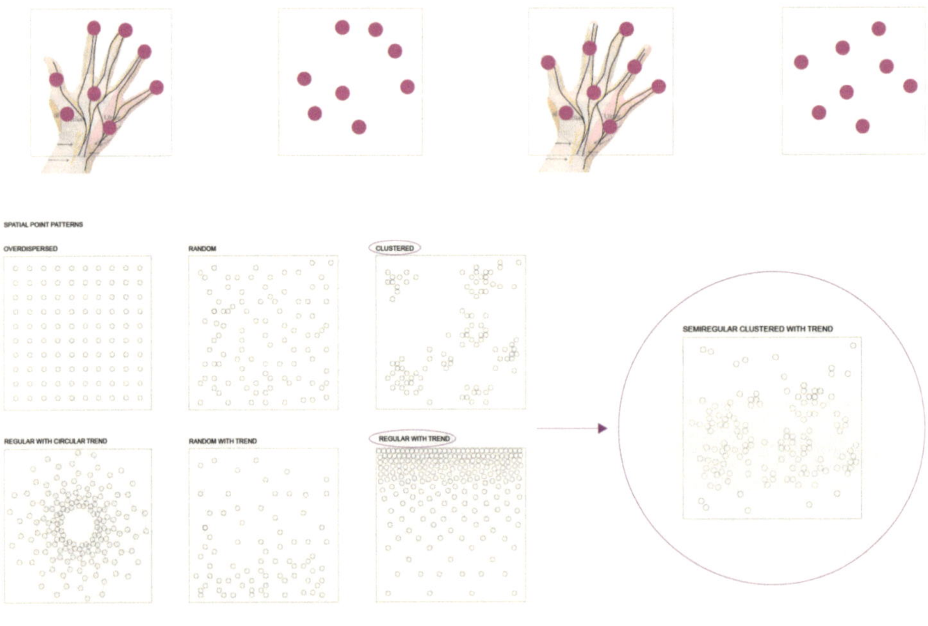

Figure 13.
Pattern creation based on trigger points inside a human hand combined with spatial point arrangements © A. Cazan/ILEK.

To achieve a sleek, canvas-like appearance, an aluminum tendering frame from Roho is utilized to secure and prestress the outer membrane also around the corners. This frame is raised 8 cm above the ground, creating a hovering effect that gives the illusion of the façade smoothly floating above the concrete platform.

The outer membrane protects the inside space and the electrical components. It consists of a silver PVC-coated PES membrane onto which round permanent neodymium magnets are positioned: these measure 15 mm in diameter and 2–3 mm in thickness. By placing one magnet on the inside and one on the outside of the membrane, the connection is made solely by the electromagnetic field, making a later dismantling, reuse, and even repositioning of magnets extremely simple.

On the inside of the façade system, there is an additional layer made of highly flexible elastane with iridescent visual properties, onto which round permanent magnets with the same 15 mm diameter but only half the thickness (1, 5 mm) are placed (see **Figure 14**).

This configuration allows for a dynamic interaction between the inner and outer layers, enhancing the overall sensory experience and interaction. While the outer membrane was completed in May 2023, a mock-up of the inner layer has been temporarily installed. It offers visitors a chance to experience the different haptic qualities and to test interaction scenarios between inside and outside space.

4.4 System dynamics

By touching the inner or outer side of the façade, users push that specific area of the textile back toward the core of the system, thus triggering an interaction. This inward movement is continuously monitored by ultrasonic sensors, which measure the distance to the default state of the membrane. The sensors then transmit this information to the Arduino microcontroller, which processes the data and sends corresponding commands to the appropriate actuators (**Figure 15**). For each sensor in the system, there is an associated actuator, which is turned on only when the predefined conditions stipulated in the Arduino code are being met. Distance and time are the defining parameters in reaching the desired effect. By experimenting with the time intervals between activations, changes in polarity, and the natural vibration frequency of the membrane, different pulsation rhythms can be achieved.

Figure 14.
Prototype of the inner layer of MagneticSKIN © U. Regenscheit.

Figure 15.
Part of the interaction system seen from the inside © A. Cazan/ILEK.

The name *MagneticSKIN* derives from the use of electromagnetic properties to activate the double-layer textile skin. 24 V powered electromagnets, each capable of generating an attraction force of 800 N, are being used as actuators. By placing permanent magnets on the membrane, the electromagnets behind the membrane are able to attract and repel the latter at predefined time intervals, thereby creating a puls-like sensation on the surface. The intensity of the effect directly correlates with the number of activation points perceived by the ultrasonic sensors and the depth of the inward movement. The greater the number of active points, the more intense the overall effect, which can be seen and (more importantly) also be felt by the users interacting with it. Thus, a new form of non-verbal haptic communication is made possible, connecting users with the outer layer of the built environment and fostering interaction and communication among the users themselves.

4.5 Output and future perspectives

The feedback from the users who have interacted with the system since May 2023 has been overwhelmingly positive, with many describing a puls-like sensation similar to a heartbeat and expressing a willingness to engage with it, thus confirming the relevance of the interaction with building skins and the perception of the built environment. By having had the opportunity to observe the system in use, it can be stated that incorporating interactive elements in the façade design encourages user engagement and instills a sense of ownership over the building.

Embracing interactive technologies and haptic qualities of materials could give architects the opportunity to create immersive experiences, where architecture transcends its traditional role and becomes a dynamic medium for human interaction and sensory perception. Moreover, the research on *MagneticSKIN* and its successful implementation serves as a significant step forward in understanding the potential of haptic communication in architecture, paving the way for the creation of more immersive and user-oriented built environments in the future.

5. Outlook

The three façade systems described in the present article show the high range of functionalities that can be achieved by using textile skin systems in new ways. Due to their lightweight and flexibility, it is easy to move and deform such panels. Thus, they can easily be adapted to different architectural purposes. Moreover, extended production ranges (as shown e.g. at the multi-layered 3D-textile for *HydroSKIN* or the customized non-woven planar mesh for *FiberSKIN*) allow for new fields of application to be explored. This fits very well in the attempt to design innovative adaptive systems, that react to different conditions in a dynamic and effective way. Modern textile skins can be also designed and built in such a way that each component has a low carbon footprint and is fully recyclable. Moreover, the experimental interdisciplinary setup of the CRC 1244 allows not only to explore the potential of new systems, geometries, and functions but also to validate experimentally the developed solutions in a real-world condition and showcase the quality of the application in full scale. Currently, the first applications clad two floors of D1244 (out of 12). Additional cladding systems are to be built in the coming months. One of the next textile façade systems is called *KineticSKIN* and will be installed on the second level of the building (see **Figure 16**).

In general, adaptive kinetic façades are designed to respond in real-time to changing environmental conditions and indoor comfort requirements by means of kinetic mechanisms that allow them to dynamically adjust their form, position, or transparency. Such façade systems can allow for proper shading and enhance occupant comfort while improving energy efficiency [23, 24].

The objective of this research is the optimization of indoor daylighting conditions and the reduction of solar heat gain as well as unwanted solar radiation in the urban canyon [25]. Excess solar radiation is reflected into the atmosphere, reducing the urban heat island effect. This is performed by reorienting the wings (façade modules) in response to changing weather conditions. The upper wing tracks the sun and reflects solar radiation to its source, thereby reducing undesired solar heat gain

Figure 16.
Rendering of KineticSKIN to be installed at the D1244. © M. Jeong/ILEK.

and reducing building energy consumption. At the same time, the lower wing serves users' visual comfort (illuminance and view).

A small number of actuators were employed to prevent adding weight to the façade and to keep energy usage low. During the hot summer months, the façade can minimize solar heat gain by shading the windows, thus reducing the demand for air-conditioning. In winter, it can allow sunlight to naturally warm up the interior, minimizing the need for heating. The system is designed to optimize natural daylight penetration into a building's interior so that artificial lighting needs are reduced. Moreover, it enhances comfort by regulating indoor temperatures and reducing glare.

PAOSS represents another approach of a targeted, kinetic sun and glare protection system using a simple, resilient, and low-energy actuation mechanism. The pneumatically operated origami sun shading system-abbreviated "PAOSS"—is used for the targeted control of light transmission. It combines the esthetic and material-immanent qualities of textile materials with the functional aspects of integrated active pneumatic actuators to initiate the change of shape e.g. to open the elements (**Figure 17**). Textile folding structures are particularly suitable for changing their shape from a large shading area to a minimal folded state and vice versa by reversible folding. They are therefore highly interesting as selective sun and glare protection elements for improving user comfort and reducing energy consumption. The National Aeronautics and Space Administration (NASA) has developed an origami folding geometry for astrophysical purposes called "Starshade" [26], which is characterized by a particularly large difference in area between the opened and folded closed state. An adaptive, pneumatically actuated sun and glare protection system inspired by "Starshade" was designed and developed to be embedded as an interlayer in ETFE cushion façades. Through the use of active components, it is possible to achieve a targeted, partial, or full-surface regulation of light and radiation transmission, as well as the back-reflection properties of the façade [27]. The ETFE façade is planned to be installed on the eleventh floor of D1244.

Within three to four years all 12 floors of D1244 will be clad with different adaptive façade systems. One of the focuses will be set on insulated glass units, integrating further functions such as cooling, energy harvesting and storage, etc. As soon as all the façades are installed, the next cycle will start, thus establishing for the D1244 the experimental character of a laboratory at a real scale.

Figure 17.
Visualization of PAOSS in un-activated closed state (right), targeted partial glare protection state (middle), and full-surface shading state (left) © C. Eisenbarth/ILEK.

Acknowledgements

The Collaborative Research Center CRC 1244 has been funded by the German Research Foundation (DFG)-Project-ID 279064222–SFB 1244. The five described projects have been made possible through the research work of different interdisciplinary teams and thanks to the support of several industrial partners and the engagement of the ILEK technicians (T. Tronsberg and M. Berndt). The authors are grateful for the support.

HydroSKIN Team: C. Eisenbarth, W. Haase, L. Blandini, W. Sobek (ILEK)

Partners: Dr. Zwissler Holding AG, Essedea GmbH & Co. KG

FiberSKIN Team: M. Jeong, L. Blandini, J. Lopez, F. Kokud, M. Matheou (ILEK)/M. Voigt, D. Roth (IKTD)

Partners: Sailmaker International (i-Mesh), Hörmann KG Verkaufsgesellschaft, TRUMPF Werkzeugmaschinen SE + Co. KG

MagneticSKIN Team: A. Cazan, L. Blandini, H. Raisch, F. Kokud (ILEK)

Partners: Roho GmbH, Koch Membranen GmbH, Mehler-Texnologies GmbH

KineticSKIN Team: M. Jeong, M. Matheou (ILEK)

Partners: Josef Gartner GmbH

PAOSS Team: C. Eisenbarth, W. Haase, Y. Klett, L. Blandini, W. Sobek (ILEK)

Partners: Global Safety Textiles GmbH, Carl Stahl AG, Mehler Texnologies GmbH

Video materials

Video materials referenced in this chapter can be downloaded at: https://bit.ly/46n1DgX

Author details

Lucio Blandini[1*], Christina Eisenbarth[1], Walter Haase[1], Moon-Young Jeong[1], Michael Voigt[2], Daniel Roth[2], Arina Cazan[1] and Maria Matheou[1]

1 Institute for Lightweight Structures and Conceptual Design (ILEK), University of Stuttgart, Germany

2 Institute for Engineering Design and Industrial Design (IKTD), University of Stuttgart, Germany

*Address all correspondence to: lucio.blandini@ilek.uni-stuttgart.de

References

[1] UNEP. 2020 Global Status Report for Buildings and Construction: Towards a Zero-469 Emissions, Efficient and Resilient Buildings and Construction Sector. Nairobi: UNEP; 2020

[2] Blandini L. Glass facades: Present and future challenges. In: Weller B, Schneider H, Louter C, Tasche S, editors. Engineered Transparency 497 2021. Berlin: Ernst & Sohn; 2021. pp. 1-12

[3] Bedon C, Honfi D, Machalická KV, Eliášová M, et al. Structural characterisation of adaptive facades in Europe—490 part I: Insight on classification rules, performance metrics and design methods. Journal of Building Engineering. 2019;**25**:100721. DOI: 10.1016/j.jobe.2019.02.013

[4] Blandini L, Haase W, Weidner S, Böhm M, et al. D1244: Design and construction of the first adaptive high-rise experimental building. Frontiers in Built Environment. 2022;**8**:814911. DOI: 10.3389/fbuil.2022.814911

[5] Intergovernmental Panel on Climate Change (IPCC). Climate change 2021: The physical science basis. In: Contribution of Working Group I to the Sixth Assessment Report of the Intergovernmental Panel on Climate Change. Cambridge: Cambridge University Press; 2021

[6] Leistner P, Kaufmann A, Koehler M, Würth M, Hofbauer WK, Dittrich S, et al. Bauphysik urbaner Oberflächen. Bauphysik. 2018;**40**:358-368. DOI: 10.1002/bapi.201800009

[7] Eisenbarth C, Haase W, Blandini L, Sobek W. Potentials of hydroactive lightweight façades for urban climate resilience. Civil Engineering Design. 2022;**4**:14-24. DOI: 10.1002/cend.202200003

[8] Eisenbarth C, Haase W, Blandini L, Sobek W. Climate-adaptive façades: An integral approach for urban rainwater and temperature management. In: Structures and Architecture A Viable Urban Perspective?. Aalborg: CRC Press; 2022

[9] Eisenbarth C, Haase W, Blandini L, Sobek W. HydroSKIN: Lightweight façade element for urban rainwater harvesting and evaporative cooling. In: Proceedings of the Facade Tectonics 2022 World Congress. Los Angeles: Facade Tectonics Institute; 2022

[10] Smith A, Gill G. Residensity-A Carbon Analysis of Residential Typologies. Chicago: Adrian Smith + Gordon Gill Architecture (ASGG); 2018

[11] Jaber S, Ajib S. Evaporative cooling as an efficient system in Mediterranean region. Applied Thermal Engineering. 2011;**31**:2590-2596

[12] Szilágyi A, Farkas I, Seres I. Evaporation cooling with ceramics in air-conditioning system. Journal of Scientific and Engineering Research. 2018;**5**(11):152-157

[13] Pires L, Silva PD, Castro Gomes JP. Performance of textile and building materials for a particular evaporative cooling purpose. Experimental Thermal and Fluid Science. 2011;**35**:670-675. DOI: 10.1016/j.expthermflusci.2010.12.017

[14] Schmid FC, Haase W, Sobek W. Textile and film based building envelopes-lightweight and adaptive.

Journal of the International Association for Shell and Spatial Structures. 2015;**56**:61-74

[15] Rentz A, Oei M, Eisenbarth C, Haase W, Böhm M, Blandini L, et al. A hydroactive facade for rainwater harvesting and evaporative cooling: Dynamic modeling and simplification for application in optimization-based long-term building operation strategy. In: Proceedings of the Conference on Control Technology and Applications (CCTA 2022). Trieste: IEEE; 2022

[16] Khan FR. Appendix I-Current trends in concrete high-rise buildings. In: Coull A, Smith BS, editors. Tall Buildings. Pergamon: University of Southampton; 1967. pp. 571-590. DOI: 10.1016/B978-0-08-011692-1.50033-X

[17] Blandini L, Eisenbarth C, Jeong MY. Adaptive Facade Systems, Detail, Nr. 9, S. 12-15, 2023

[18] Voigt M, Chwalek K, Roth D, Kreimeyer M, Blandini L. The integrated design process of adaptive façades-A comprehensive perspective. Journal of Building Engineering. 2023;**67**:106043

[19] Hellfritz H. Innovationen Via Galeriemethode. Königstein im Taunus: Eigenverlag; 1978

[20] Attia S, Lioure R, Declaude Q. Future trends and main concepts of adaptive facade systems. Energy Science & Engineering. 2020:**8**(9):3255-3272. DOI: 10.1002/ese3.725

[21] Pallasmaa J. The Eyes of the Skin: Architecture and the Senses. Chichester, Hoboken, NJ: Wiley-Academy, John Wiley & Sons; 2012

[22] Fadzil A. Sensory Immersion in Architecture. Portsmouth: University of Portsmouth; 2015

[23] Aksamija A. Sustainable Facades: Design Methods for High-Performance Building Envelopes. New Jersey: John Wiley & Sons; 2013

[24] Schnittich C, Krippner R, Lang W. Detail: Building Skins. Basel: Birkhäuser; 2006

[25] Jeong MY, Matheou M, Blandini L. Optimisation of daylighting performance through adaptive kinetic envelopes. In: Kanaani M, editor. The Routledge Companion to Ecological Design Thinking Healthful Ecotopian Visions for Architecture and Urbanism. New York: Routledge; 2022. pp. 251-262

[26] Sigel D, Trease BP, Thomson MW, et al. Application of origami in starshade spacecraft blanket design. In: ASME 2014 Design Engineering Technical Conferences and Computers and Information in Engineering Conference, Buffalo, New York. 2014. DOI: 10.1115/DETC2014-34315

[27] Eisenbarth C, Haase W, Klett Y, Blandini L, Sobek W. PAOSS: Pneumatically actuated origami sun shading. JFDE. 2021;**9**:147-162. DOI: 10.7480/jfde.2021.1.5535

Chapter 6

Smart Façades: Technological Innovations in Dynamic and Advanced Glazed Building Skins for Energy Saving

Silvia Brunoro and Valentina Frighi

Abstract

This chapter deals with the analysis of the potential offered by the integration of smart solutions in dynamic glass façades to improve buildings' energy performances. Dynamic solutions are here examined with reference to dry ventilated systems, active and passive cooling, solar gain, greenhouse effect, and technologies able to react and self-regulate, according to the environmental inputs. The first part is dedicated to the state of knowledge, assessing the performance evolution of dynamic and interactive architectural envelopes (smart skins). Then, the core of the chapter is divided into clusters according to different strategies that allow the building skin to react and self-regulate according to the environmental inputs: double-layer glass façades, solar shadings, PV integration, etc. The goal is to produce a sort of "smart skin guideline" divided by requirements/strategies of intervention to investigate a range of solutions able to regulate buildings' behavior and characterize their image: from systems that allow to transform solar gain into heat to improve buildings' energy performance in winter season, to others that integrate passive cooling, to systems that transform the façades in a real active element of energy production, thanks to the integration of renewable energy sources.

Keywords: glass building skin, smart envelope, double layer façades, active and passive system, advanced building skin, smart windows

1. Introduction

The aim of this chapter is to explore the innovative incorporation of glass into façade systems to promote the energy efficiency and enhance the architectural perception of buildings. The research is focused on the sustainability of glass as a material for façade, as a very powerful opportunity, often associated with natural light, and lightness, and other qualities that have earned universal reception from modern architecture.

Firstly, a brief introduction frames the research background, the concept of glass envelope, its evolution, and role toward the definition of dynamic and smart envelope.

Then, the methodology is explained in paragraph 2; later technical solutions are classified by their technical requirements and energy performance. The core of the chapter is divided into clusters according to different strategies that allow the building skin to become an active and dynamic layer: double layer glass façades, solar shadings, PV integration, etc.

Finally, the goal is to produce a sort of "smart skin guideline" divided by requirements/strategies of intervention to investigate a range of solutions able to regulate buildings' behavior and characterize their image.

Historically, the use of glass envelopes was mainly focused on esthetics, as it was estimated that it did not need to be ecologically responsive to the environment.

Adverse energy and mechanical performances usually associate with excessive thermal gain and direct sunlight, have created uncomfortable buildings, and caused inefficient energy consumption [1, 2].

Hence, the application of a glazing system cannot be followed without truly understanding the underlying principles and implications.

Since energy costs have been affordable (before the oil crisis of the 1970s), the low thermal performances of the fully glazed building have been compensated by totally mechanical heating and cooling systems. By the 1970, the high costs of fuel led the building industry to develop new and performing glass products such as photosensitive and photochromic glass, and new coatings such as reflective or selective (Low-E) to help in reducing energy consumption in buildings with large glass area [3, 4].

The increasingly frequent use of transparent surfaces for the construction of building development began in the nineteenth century, during the Industrial Revolution, and involves the research and development of new materials able to guarantee energy performance like massive walls [5].

The envelope becomes progressively independent from the load-bearing structure of the building and its first requirements are to regulate the energy flows such as heat transfer, light transmission, protection of solar radiation (**Figure 1**) [6].

In the recent years, sustainability has become a more and more important feature in architecture: A sustainable design process can produce high-performance buildings that are energy efficient, healthy, and economically feasible, wisely using renewable resources to minimize the impact on the environment and to reduce, as much as possible, the energy demand [7–9].

Following the developments in international environmental policy, after the Kyoto agreement on climate change [10], the International Energy Agency has developed a set of scenarios on international energy development up to 2030 and 2050, showing how the construction sector remains, alongside the industrial sector, the most responsible for energy consumption and CO_2 emissions [11].

Consequently, several European standards and regulations concerning energy efficiency in buildings have been promulgated, focusing on the importance of the energetic control of buildings to reach, since 2020, NZEB requirements for new and relevant refurbishment actions [12].

Since the publication of the European Directive 31/2010 UE, the building envelope has been the subject of a great number of research aimed at demonstrating the possibility of build with zero or passive emission house [13].

Among the European research in recent years contributing to the envelope evolution, through experimentation of new components and materials characterized by high performance, it can be cited: Best practice for double-skin Façades (BESTFACADE), European High Quality Low Energy Buildings (EULEB), Building Advanced Ventilation Technological examples (BUILDING ADVent).

Figure 1.
GSW Headquarters, Berlin. Architects: Sauerbruch & Hutton. One of the first examples of high-efficient double layer glass façade. All moveable curtains of the façades can be controlled by the individual occupant but are also operated by a central building management system. This individual control provides for a continuously changing appearance, especially of the west façade.

As remarked in Ref. [14], for smart glass façades, the thermal and insulation properties, in addition to other important requirements like transparency to allow solar gains or solar control, are relevant aspects that are strictly linked on climate conditions and the desirable level of comfort.

Nowadays, the standardized prefabrication of high-quality multifunctional façades represents a foreseen frontier in the improvement of the envelope's performance. Dry-mounted prefabricated façades reduce the on-site construction costs and time by incorporating numerous functionalities into the same component, thanks to the inclusion of smart materials. This approach also integrates both solar active and passive solutions and optimizes building equipment, including heating, cooling, and ventilation.

Therefore, glass components' system becomes a multifunctional building layer, not only to be physically and functionally "integrated" in the building, but also to be used as an innovative chance for the building envelope design. For this reason, the façade system plays an important role for achieving energy and environmental goals [15].

Architecture of light envelopes is made of structural skins just like a shell, the first significant experiment took place in Northern Europe such as in the design experiences of Sauerbruch & Hutton, Herzog & de Meuron, Jean Nouvel, etc. [16].

In the design of dynamic envelope systems, it is fundamental to analyze the climatic and environmental conditions, to reach an adequate balance between climatic parameters, inner thermal and hygrometric conditions, and technical components/solutions.

Starting from this point, the chapter proposes an analysis of the most relevant active technologies based on their performances.

Relying on the most efficient products in terms of energy efficiency of glass components for façades, the aim of this work is to perform a multi-criteria analysis based on different requirement categories, to compare several technological possibilities for façade in terms of technological, architectural, energetic, and environmental requirements.

2. The role of glazed components

While glazed elements are an important aspect of architectural design, they are regarded as the least energy-efficient elements of the building envelope [17] since they contribute to approximately 60% of the overall energy usage of buildings [18]. In contrast to insulated walls of equivalent size, the heat-gain occurrences on windows can result in effects that are many times more impactful [19], resulting in significant implications for buildings' lighting, heating, and cooling needs.

The challenges associated with glazed elements lead to winter heat losses, due to air leaks and insufficient insulation. Similarly, in summer, they can lead to overheating due to the entering of solar radiation, which significantly elevates indoor temperatures. Additionally, beyond their architectural and energy-related aspects, these components must also fulfill essential structural prerequisites. These requirements encompass structural integrity, usability, longevity, resilience, and fire performance [20–22], particularly if considering systems with adaptive features.

Hence, effectively intervening in the design of glazing systems offers a significant chance for the construction industry to manage energy needs, contributing to the advancement of the objectives set by the European energy agenda.

2.1 Characteristics of glass products

Throughout years, artists and architects have worked with glass due to its shaping, tactile qualities, and interaction with light, also considering its stability, waterproofing, and see-through nature [23].

Since the development of the float glass technique in the 1950s, glass surfaces appropriate for construction purposes are primarily made of silica (SiO_2). Transparent typical glasses have a light-transmission coefficient ranging from 60 to 80% for wavelengths approximately falling within the 400- to 2500-nm range. Nevertheless, by altering the chemical composition of the glass through adjustments in its mixture, it becomes possible to alter this threshold or impact other aspects like its chemical, physical, or mechanical characteristics.

Due to the evolution of technological advancements and industrial methods, glass is nowadays available in a variety of shapes and compositions, typically distinguishable based on the production techniques that generated them.

2.2 Glass energy performance features

The primary characteristic of glass is its ability to transmit light, facilitated by its transparency, which arises from the interactions between light photons and the atomic structure of the glass. Generally, a glazed surface transmits most of the incoming solar radiation, depending on the constituents that form it and/or any surface modifications it undergoes.

Within the entire solar spectrum, three segments significantly impact the comfort of indoor environments in buildings, as they pass through the glass: Ultraviolet (UV), Visible Light (VL), and Infrared Energy (IR). UV light can be further categorized into three groups; two of them are rejected by the earth's atmosphere and float glass as well.

IR constitutes the heat energy emitted by the sun that enters an interior space. Managing this transmission involves limiting heat within rooms, thus averting potential overheating during the summer. The greater the ability of a glass panel to block IR, the more energy and cost can be saved.

Lastly, VL represents the portion of light visible to the human eye and encompasses natural daylight. It can contribute to undesired glare and strain on the eyes. Generally, the VL of a glass pane can be decreased by adjusting tints.

Regarding VL, glass is almost completely transparent, whereas concerning IR, it behaves opaquely; this is the cause of the so-called greenhouse effect, due to which, bodies located in a space protected by glass surfaces experience temperature elevation due to direct exposure to radiation. This energy is then re-emitted as sensible heat in the form of infrared radiation, which remains confined within the space.

IR solar radiation upon a glass surface can be reflected, transmitted, or absorbed. In normal incidence conditions, the reflected energy is the quantity of solar radiation bounced back into the atmosphere. The transmitted energy represents the solar radiation directly passing through the glass's surface, while the absorbed energy expresses the quantity of solar radiation absorbed by the glass, leading to an increase in its temperature.

Total transmission instead represents the total amount of solar radiation that, in normal incidence conditions, is transmitted through the glass. This measurement encompasses direct transmission (short-wave component) as well as the component dissipated inward due to radiation and convection (long-wave component).

The ability of a surface to absorb or emit electromagnetic radiation is reflected by the emissivity value. By its nature, glass has a high emissivity. Coatings able to act on this feature are existing, reflecting heat inside buildings and thus reducing heat losses through windows, increasing in this way the system's U-value.

2.2.1 Reference parameter for glass energy performance evaluation in buildings

Defining and characterizing the energy performance of transparent components is quite tricky due to their interaction with solar radiation. Because of these complexities, such building elements must adhere to specific criteria and be described using both thermal and optical parameters. These parameters address the system's thermal insulation capacity and the ability in regulating the quantity of solar energy into rooms. The benchmarks commonly used for assessing the performance of glazing include the following:

- Thermal transmittance, or Heat Transfer Coefficient (also called U-factor), which is the rate at which a transparent pane conducts non-solar heat flow. A lower U-factor indicates higher energy efficiency.

- Solar Heat Gain Coefficient (SHGC), or g-value. This quantifies the amount of incoming solar radiation that a window permits to pass through, either by direct transmission or absorption, and then released as heat within an indoor space. A

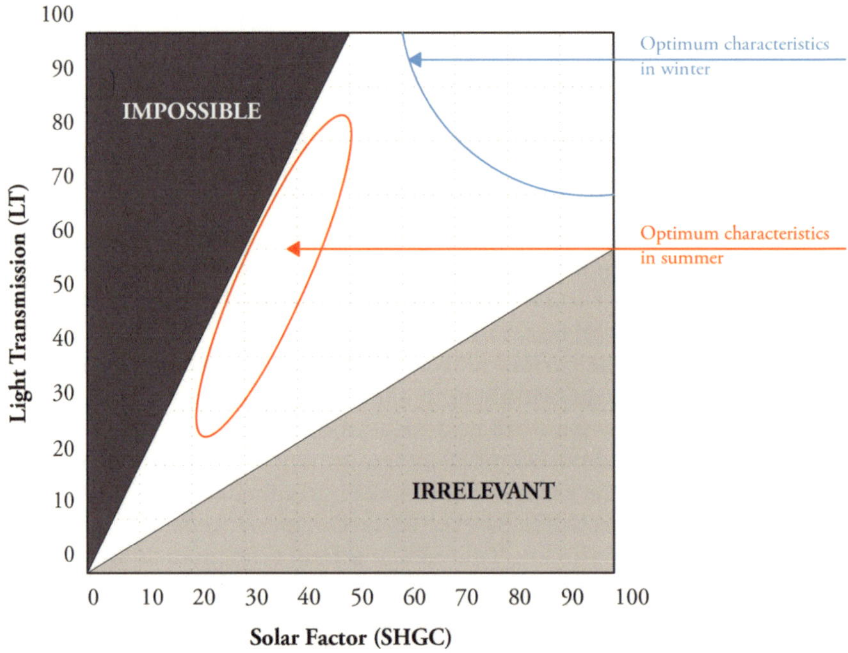

Figure 2.
Correlation between the Solar Factor (SHGC) and the Light Transmission (LT) coefficient.

lower SHGC means reduced solar heat transmission and enhanced shading capabilities. Glazed elements with high SHGC effectively capture solar heat in winter, while systems with low SHGC are efficient at reducing cooling loads in summer.

- Visible Light Transmittance, or light-transmission coefficient (τl), is the fraction of the visible sunlight spectrum perceived by the human eye which passes through the glazing. It is represented by a value between 0 and 1, with higher values denoting more transmitted visible light.

An additional crucial parameter for assessing the energy performance of transparent systems is the Light to Solar Gain (LSG) ratio, which measures the spectral selectivity of a glass system as an expression of the efficiency of different glass types in transmitting daylight (the visible solar radiation) while blocking heat gains. It is calculated as the ratio of VLT to SHGC; a higher LSG means more daylight transmitted without adding excessive heat. This parameter becomes particularly important for daylighting, especially in summer when more light is desired with minimal solar gain. Therefore, a higher LSG number implies a brighter room without excessive heat accumulation (**Figure 2**).

3. Methodology: Categories of requirements and performances

This work focuses on the technological innovation in glass façades systems and investigates the most relevant aspects from architectural and energetic point of view.

The first analysis involves identifying relevant factors—presented in the form of criteria—that define the main features of each system.

1. Architectural features (esthetical and functional):
Esthetical aspects, façade integration, transparency, translucence, multimedia.
2. Energy properties and comfort
Energy issues and aspects related to human comfort: light transmission, solar factor, thermal insulation, greenhouse effect (solar gain), thermal inertia, acoustic insulation, natural/mechanical ventilation, solar device integration, renewable energy integration.
3. Environmental properties
Environmental aspects: ecological and sustainable features (recyclability, green products).
4. Economical features

Table 1.
Significant criteria for façade components.

Each macro-category of requirements (architectural, energy, environmental, and economic) is further divided into sub-categories to define the peculiar aspects of components under investigation, as well as their operational strategies, in terms of the environmental and dynamic benefits within the overall concept of the building.

The assessment methodology for the analysis of the architectural and technological characteristics of the façade systems is based on the categories and sub-categories of requirements listed below (**Table 1**). These have been defined by authors based on their significance concerning the whole building envelope system—in relation to both technical and aesthetic features—and parameters indexed in the regulation framework for transparent building components, plus other qualitative considerations made based on their description related to the integration with technical systems and cost-related aspects.

According to this criteria, three macro-categories of environmental strategies have been proposed, to classify, in the following paragraphs, the most relevant technological innovations in advanced glass building skins for energy saving:

- STRATEGY 1—Passive solar gain: double layer glass façades.
- STRATEGY 2—Summer overheating control.
- STRATEGY 3—Renewable solar energy—semitransparent PV, bio-adaptive glass.

The purpose is to outline criteria and operational tools to guide and inform the design of innovative envelopes, allowing targeted choices to be made in relation to the foreseen interventions, to obtain the desirable levels of quality.

Finally, a technological multi-criteria matrix is defined, in which the façade solutions are briefly described and compared by crossing the strategies of interventions with the requirements described in **Tables A1–A3**.

3.1 Passive solar heat gain: double layer glass façades

Solar gain holds significant importance in cold and temperate climates, as it plays a crucial role in reducing heat losses and harnessing passive solar incidence, thus contributing to the overall energy balance. Historical solar passive design—e.g., the

"Trombe wall"—can be considered as a precursor to modern double skin systems [24].

A double Skin Façade is an advanced building skin, originally born in Northern Europe, that can dynamically respond to varying ambient conditions, able to:

- maximize daylighting with integral solar heat gain control;
- improve heat exchange by greenhouse effect in the buffer zone; overheating, as a minor problem, is solved by blinds and buffer zone;
- reducing air-conditioning loads;
- allow natural ventilation (**Figure 3**).

A typical double glass envelope system comprises a layer of single glass and a layer of double-glazing, separated by an air space that can incorporate a range of integrated sun-shading, natural ventilation, and thermal insulation devices or strategies. During the winter season, a double skin façade can improve heat gains coming from solar radiation by means of the greenhouse effect and reduce the heat losses due to the slow air flow that can lower the heat transfer from inside to outside. Glass panes maintain a warm surface temperature on the inside, enabling more effective utilization of the space close to the window [14].

In winter, the air cavity, heated by the sun, becomes a warm buffer zone, while in summer the stack effect of fresh air (passive ventilation) removes the exceeding heat [25].

In addition to energy efficiency and the U-value of the glass, other important requirements include acoustic control, water penetration resistance, and daylight control, which are crucial for ensuring office building comfort.

In summer, control of solar heat gain is ensured through shading devices placed within the air cavity. Additionally, the cavity itself can help dissipate some of the incoming solar radiation through the passive ventilation effect. Various configurations for shading devices exist; they can either be fixed elements or, typically, operable units that are either controlled by the occupant or by sensors within the building [26].

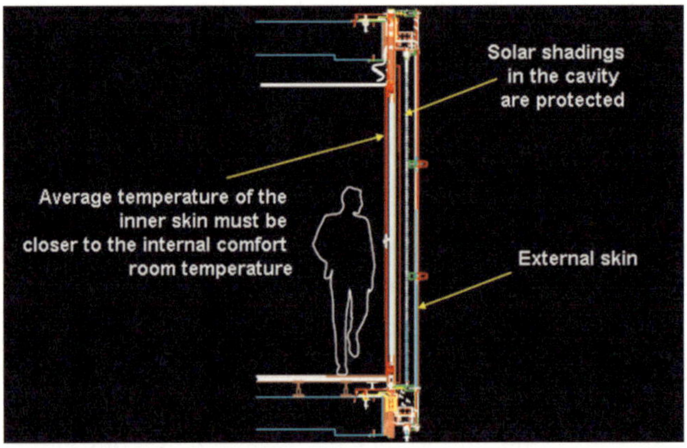

Figure 3.
Conceptual scheme of a double layer glass façade.

In Mediterranean Countries, solar gains control in building design that is relevant; therefore, double skin façades may lead to overheating during summer months if there is no appropriate façade design, ventilation technique building orientation, and provision of shading [27].

The use of improved solar U values for glazing favors the absorption and reflection of heat, to minimize solar heat gain. This can be accomplished using spectrally selective glazing, which can selectively respond to different wavelengths of solar energy. It allows visible light to pass through while rejecting unwanted invisible infrared heat [28]. There are products available on the market that have successfully achieved this characteristic, permitting for much clearer glass than previously available for solar control glazing, as reported in paragraph 3.

In addition, complete glass façades need an accurate daylighting control to avoid excessive glare and heat. This is important primarily for two reasons: firstly, it reduces the amount of electrical lighting required, and secondly, because the quality of light from daylight is preferential to electrical lighting. It is proved that the health and productivity of office personnel are highly influenced by the quality of lighting in the environmental workplace; natural light is particularly relevant for well-being and mental health of the occupants [28].

A classification system to assess and compare different kinds of double skin façades is proposed, by considering their main characteristics regarding building construction and their environmental behavior, related to the ventilation system.

According to these parameters, a categorization can be done by considering different typologies of construction and cavity ventilation [29].

Based on their construction features, façades can be classified by full height, corridors, or cells, depending on the air cavity dimension and division. An appropriate design of the air space is crucial to the double façade.

In full height façades, the façade is undivided from the bottom to the top, creating a unique air cavity that benefits from the stack effect. On warm days, hot air collects at the top of the air space. In certain cases, the undivided air space can be a big atrium, in which people can stay and enjoy this "environmentally dynamic interstitial space" [30] that can be used for spaces with low occupancy (meeting rooms or cafeterias). Sometimes, vegetation is used instead of traditional shading devices, to improve air quality and well-being [31].

Corridors façades are divided by floor, best for fire protection, heat, and sound transmission, or divided vertically into bays to optimize the stack effect.

This kind of construction is commonly used when it is needed to have fresh air and exhaust intakes on every floor, allowing for maximum air changes. When the cavity is divided into vertical corridors, air flows across the façade through openings, allowing for better natural ventilation.

A third typology consists of high prefabrication elements—one floor high—ready to be installed, that are called cells. Double layer glass cell façades have a floor-by-floor divisions that allow a construction speed by a simple repeating unit, which can maximize ventilation thanks to fresh air and exhaust intakes on every floor. Moreover, the compartmentalization of the air cavity divided into small zones improves fire security, noise requirements, and heat transfer from one section to another (**Figure 4**).

Another classification can be made according to ventilation.

When ventilation is natural, the air entered from the bottom, warms thanks to the sun radiation across the glass panes. Hot air rises according to natural physics principals until openings at the top of the cavity allowing expelling it out and replacing it by

Figure 4.
Different types of double glass skin. From left to right: full height, corridors, cells.

cooler air drawn in from the outside. Sometimes, especially in hot climates—when there is the risk of low pressure and lack of stack effect—the offices on the top floors can suffer from overheating due to the accumulation of hot air in the cavity adjacent to their space. For this reason, it is a good practice to insert mechanical air vents to assure the hot air will be discharged [32].

In this configuration, the single-glazed outer skin is used primarily for moderate to extreme temperature within the façade and for the protection of the air cavity contents (e.g., shading devices). External single glass is generally safety or laminated glass, while the internal skin offers the insulating properties to minimize heat loss (double or triple glass) [33].

Natural ventilation includes the necessity of openings in the outer skin (moveable glass panes or grids), while windows on the interior façade can be opened or not. Ventilation openings in the inner skin allow the building's users access to airflow that can be used to cool and ventilate the space.

The exterior glazing of the double skin creates a layer that in most case can be accessible by the inhabitants for natural ventilation: This buffer zone is a key component of the naturally ventilated double skin façade, typically 60-90 cm of thickness, and can include plants and vegetation to mitigate the temperature.

Furthermore, the use of internal moveable windows can allow for night-time cooling of the interior thereby lessening cooling loads of the building's HVAC system.

In double layer glass façades with forced ventilation, the air space between the two layers of glazing becomes part of the HVAC system. In this configuration, the thermopane units (optimal U-value) are placed on the external of the main façade of double glazing. In winter, the heated air between the glazing layers flows through the cavity and the outer layer of insulating glass minimizes heat-transmission loss. In summer, fresh air is supplied by HVAC through the cavity and contributes to the façade cooling.

Numerous studies prove that double skin façade presents many advantages over a conventional—single skin—façade [34–36]. A double skin dynamic envelope in a cold temperate climate, like the United Kingdom, can reduce energy consumptions by 65%, running costs by 65%, and cut CO_2 emissions by 50%, compared to a single skin building. The same study reports that cost exercises of buildings employing a double skin façade may cost as little as 2.5% based on gross internal floor area [37].

Regarding sustainable design, the double layer façade offers strategies of solar heat gain control, increased daylight, moderation of temperature, and natural ventilation strategies to improve the thermal behaviour of the building. The ability to engage and control these environmental aspects inevitably leads to increased energy efficiency and improved occupants' comfort.

Significant studies in sustainable architectural design have shown that construction cost of double layer glass façades can be two or three times higher than a traditional curtain wall, anyway double skin buildings can significantly reduce overall long-term operating/energy costs making the increased initial capital costs justifiable (affordable). The goal of these systems is not only to be environmentally responsible but also to greatly improve working conditions for the occupants of these buildings through access to day lighting, natural ventilation, and greater control over the workplace atmosphere. Moreover, a double skin system also offers a choice for renovation of existing building façades to transform into more energy efficiency buildings and improve the architectural value.

3.2 Strategies to control summer overheating

Over time, glass, a material that once performed only the function to allow the entrance of light, has evolved into a dynamic filter, capable of assuming a more interactive role in managing the internal environment of buildings.

Various experimental and groundbreaking glazed systems, encompassing a broad array of functions and applications, are currently accessible in the market or, at the very least, in the developmental phase, having different features from many points of view.

This group includes what is referred to as "dynamic glazing," fenestration products capable of altering their performance attributes by modifying their characteristics of transparency, gloss, coloration, and solar radiation screen while maintaining the structural properties of the glass [38].

Dynamic glasses, otherwise defined as "smart windows," can be controlled through a variety of means, enabling end-users to govern the interactions between external conditions and the internal environment. This capability results in the creation of a smart, adaptive building envelope that contributes to cost savings related to heating, cooling, and lighting.

Therefore, the massive adoption of these innovations holds the potential for substantial decreases in energy usage and, by obviating the need for supplementary devices to regulate incoming solar radiation, also in costs, maximizing at his best natural daylight while minimizing issues like glare and heating/cooling demands.

To fully understand the potential of these glazing components, a brief description of the most promising glazing technologies has been compiled, drawing from recent literature on the subject. These technologies have been categorized based on their performance characteristics, whether they are equipped with static or active features.

3.2.1 Static performance glazed components

Static performance glazed components are technologies that act passively on the control of incident solar radiation, thus meaning without varying their performance feature over time.

An example of passive solar control is provided in glass treated with metal oxides, during the production process. This is the case of tinted glasses, in which the applied oxides allow to vary color and optical properties of glass, conferring to it the ability to absorb certain sections of incident solar radiation, thereby reducing the amount that passes through them. Normally, the coloring is achieved using ionic or phase-dispersed color substances, formed by particles' aggregates that affect the level of transparency as they lead to light diffraction and reflection due to the particles dispersed throughout the glass material. A colored glass with discrete properties should block solar energy wavelengths ranging from 800 to 2000 nm, still maintaining a reasonable level of visible transmission. Thermal transmission in a room with tinted glass can be reduced by more than 20% [39].

The application of metallic oxides in the form of coating can generate also reflective glasses, in which the pane's reflection toward near IR is increased; they are characterized by a predetermined selective reflection toward solar radiation and by a low SHGC. Compared to tinted glazing, reflective glasses have significant effects on diminishing solar transmission; this attribute makes them a favorable choice for regions characterized by climates where the reduction of solar heat gain is crucial, particularly in cooling-dominant environments.

Acting on the emissivity of the glazed panes allows to reduce heat exchanges through them, without compromising their transparency. This is the case of low-emissive glasses (Low-E), transparent toward solar thermal radiation but reflecting regarding IR radiations. A weak point of Low-E is that they multiply the effect of thermal solar radiation, increasing overheating as well; for this reason, they are particularly suitable to be applied in heating-dominate climates (rigid climatic contexts) where high solar factor (SHGC) is required, as well as a low thermal transmittance coefficient (U-factor). Employing Low-E in Insulated Glass Units (IGUs) allows to obtain thermal transmittance values between 1.7 and 1.0 W/m^2K, for double glazing, and lower than 0.7 W/m^2K for triple glazing, reducing solar heat gain through windows up to the 48% [40].

Combining the thermal insulation property of Low-E with the sun-blocking features of reflecting glasses results in selective glasses. These glasses utilize coatings to hinder the entry of infrared radiation, while maintaining controlled levels of light transmission and simultaneously restricting the solar factor's impact. These selective coatings filter the electromagnetic waves, admitting most of the incoming solar radiation in both visible and near-infrared spectra. However, they reflect long-wave radiations (far-IR) emitted by warm objects inside rooms. This approach has the effect that during the winter season, when solar rays are inclined—typically parallel to the slender glazed element—the radiations can permeate the system, triggering the greenhouse effect, exploited in passive thermal energy strategies to decrease heating demands (**Figure 5**).

Other static-performance glazing technologies are Vacuum Insulated Glasses (VIG) and TIM (Transparent Insulating Material) glasses. Both are fully available on the market but still with a quite high cost.

Figure 5.
From left to right: tinted glass at Miami International Airport; reflective glass at Mann Island Development Building; selective glass at the Blue Pavilion, Fiera del Mare, Genova.

Standard VIGs consist of two glass panes, generally of low emissive, set apart by a vacuum cavity. The benefit of this technology is the combination of an exceedingly thin profile with highly effective thermal insulation characteristics[1], due to the vacuum in the space between the glass panes, which prevents convective exchanges between them. However, they still present some shortcomings; these encompass vulnerabilities to pressures originating from wind and vibrations that could impact the glass surface, as well as the challenge of maintaining an airtight seal along the edges to prevent the reestablishment of conduction to its normal level.

TIM glazing systems instead comprise in the space between two glass' panes a transparent or translucent insulating material [19], combining diffused lighting features with very limited thermal losses[2] [43].

With their translucency (typically having a SHGC equivalent to that of a standard IGU), these materials enable the dispersion of natural light, encompassing both direct and diffused illuminations. This feature proves advantageous in spaces where there's no necessity for complete visual transparency to the external environment, effectively averting instances of glare.

TIMs are generally classified under four categories, according to the structure of the TIM layer; TIM glazing systems generally employ plastic capillaries or honeycomb structures insulating materials that can be made of polycarbonate, poly-methyl-methacrylate, or aerogel [44]. Nonetheless, a limited number of investigations have delved into the thermal and optical capabilities of glazed systems incorporating Transparent Insulation Materials (TIMs), as the majority of prior research has primarily focused on their usage within solar collectors and solar walls [45].

Another worth-mentioning technology is the one employed in Heating Glasses (HG) where an electrically connected metallic coating is introduced to the inner pane of an IGU; upon applying a low voltage, this coating heats up and directs warmth toward the interior, effectively minimizing heat loss through the transparent surface. These systems completely remove the risk of condensation between glass' panes, contributing also to the heating system of the entire building, as they generate about 0.42 Kw/m^2 of heat. Typically, heating glasses require 100–300 W/m^2 to operate, resulting in a pane temperature of roughly 40°C. If used solely for preventing condensation, their operational power can be reduced by half.

Appeared on the market in 1980, now heating glasses are produced by several companies all over the world; among the others, significant is the device produced by Vitrius Technology since it can re-calibrate and re-program itself in real time, improving its performance based on end-user needs.

3.2.2 Dynamic performance glazed components

Contrary to static performance components, dynamic performance systems, otherwise called "smart glazing," can modulate their optical characteristics according to different inputs of various natures. Smart glazing technologies can be active if they change using external signals (such as electrical direct currents) or passive if they automatically respond to environmental variations, such as air temperatures or solar radiation.

---

[1] VIG systems reach for a U-value of about 0.4 W/m^2K [41].

[2] A study [42] reveals that the presence of a TIM structure can suppress convective heat transfer through the windowpanes and cause a significant reduction in radiative heat transfer.

Active systems comprise of technologies based on Electrochromic (EC), Gasochromic (GC), Suspended Particles Devices (SPD), and Liquid Crystal Devices (LCD), while Thermochromic (TC) and Photochromic (PC) technologies are examples of passive smart glazing types.

Electrochromic glasses are systems capable of altering their light transmission characteristics, usually achieved by modifying their color and optical attributes after the application of an electric field. ECs are obtained by introducing a layer of micro-liquid crystals between two glass panes within the gap. These liquid crystals facilitate reversible electrolytic reactions that, when exposed to a potential difference, can change their coloration to the extent of becoming transparent [46].

During the transition, light transmission of the system modifies from about 1-4% (in the opaque state) to 60–63% (in the transparent state); the SHGC instead stays between 0.63 and 0.26 for the clearest state and between the 0.31 and 0.04 for the darkest state.

Once the change has occurred, no electricity is needed for maintaining the shade that has been reached.

The speed at which EC devices switch can range from a few seconds to several minutes, according to the type of technology used and the size of the window.

Most of today's available devices function in either on- or off-states only, although technologies enabling adjustable degrees of transparency can be readily implemented [47].

Recent progresses in EC materials, particularly those based on transition-metal hydrides, have resulted in the creation of hybrid systems with reflective properties. These systems shift from being absorbent to reflective, allowing them to alternate between transparent and mirror-like states. While these materials adhere to the foundational concept of conventional EC, they approach the issue differently: transitioning from a transparent state (when inactive) to a reflective state upon the application of voltage.

Additionally, the exploration of self-powered EC glazed components activated by photovoltaics (PVs) has been undertaken [48]. However, their transparency is substantially limited due to the presence of the PV layer. These devices employ sputtered titanium and tungsten oxide films as the electrochromic layer, combined with a photoactive layer composed of dye solar cells, often constructed using dye-sensitized titanium oxide (TiO2) [49].

Gasochromic glasses are obtained by introducing a gasochromic layer between two transparent panes. This layer reacts with a mixture of diluted hydrogen gas (usually combined with Argon), resulting in a color change and alteration of the system's transparency due to a catalytic reaction with the glass composition. The degree of transparency in these devices depends on the quantity of hydrogen the gasochromic layer has been exposed to. GC glasses maintain unobstructed visibility from the interior to the exterior in all operational states.

Tungsten oxide is the most used material for GC applications [50], often accompanied by a thin catalyst layer, although devices employing a thin layer of Wolfram are also accessible.

The alteration in the transmission characteristics of glass enables a reduction in both visible and overall solar energy transmittance rates of an IGU, reducing them from 0.63 and 0.49 to 0.20 and 0.17, respectively (when the interior pane is coated with a conventional low-E coating). By introducing a solar-control coating, even lower SHGC values can be achieved.

In contrast to other passive smart glazing systems, GCs require additional control equipment, such as a gas supply unit and a control unit. The gas supply unit encompasses an electrolyzer and a pump, linked *via* pipes to the glazing system in a closed-loop arrangement. Ideally, this gas supply unit is integrated within the external façade of the building. A single gas supply unit has the capacity to furnish sufficient gas to activate gasochromic glazing covering an area of 10 m^2 [51]. On the other hand, the control unit facilitates both manual and automatic regulations. When integrated into a bus system, this unit allows the glazing to be switched, optimizing lighting conditions, thermal comfort, and/or overall building energy consumption.

SPDs are electroactive devices in which the application of an AC voltage prompts particles to transition from a random arrangement to an aligned one, becoming the glazed components transparent. In the absence of an electric charge, SP windows absorb light, leading to a reduction in light transmission. The typical ranges of light transmission and SHGC—when transitioning between transparent and opaque states—are approximately 60–0.5% for VLT and 0.57–0.06% for SHGC, with switching times of some seconds.

Simulation outcomes have demonstrated that switchable SPD smart windows, when in the off and automated states, can result in a net energy reduction of up to 58% compared to double-glazing low emissive IGUs [52].

Conversely, LCDs employ materials with a bars-molecular structure that, under the influence of an applied voltage, can alter the light transmission characteristics of the systems; most of the LCDs tend to disperse light, leading to their becoming white and semi-transparent [51].

Typically, LCDs consist of a layer containing droplets dispersed within a polymer matrix. When the voltage is off, these droplets scatter light due to the disparity in refractive indices between the matrix and the droplets.

In the active state, the light transmittance of liquid crystal glazing does not exceed 70%, whereas in the inactive state, it remains around 50%. These systems diffuse direct incident solar radiation without effectively blocking it, which results in a SHGC usually ranging from 0.69 to 0.55.

In contrast to SPDs, these systems are primarily used for privacy applications [53] and need of a continuous supply of electrical energy to remain operational.

Thermochromic are glazed systems in which a temperature variation triggers a response in the material. This reaction enhances its reflective capability, making it particularly responsive to IR. Consequently, there is a modification in light absorption related to the external surface temperature, making these devices opaque when reaching a critical temperature (specific for each product).

In general, the most employed TC technology is the one that uses tungsten trioxide or vanadium dioxide (VO_2) coatings to obtain this reversible behavior [54]. VO_2 is the most common material, despite its many shortcomings as the high transition temperature, comprised between 10°C and 65°C, so much higher than indoor temperatures [55]. To address these limitations, alternative approaches involve incorporating additional substances like W and F into VO_2 to lower its transition temperature, or utilizing specialized gels placed between the two layers of plastic film.

Nevertheless, this leads to a reduction in the VLT of these systems. To counteract this effect, anti-reflective coatings are applied to increase it [46]. Another notable limitation of traditional fully passive thermochromic technology is its inability to adequately adapt transparency based on outdoor climate conditions. While it does regulate solar radiation, it overlooks the indoor temperature rise due to heat entering per convection [17]. However, these systems maintain a relatively low cost in comparison with more complex alternatives [56].

Photochromic glasses can vary their optical properties due to external light-intensity variations by means of the presence of organic or inorganic compounds that act as optical sensitizers [57]. Their conduction becomes reversible once the exposure to radiation stops; this reversibility is allowed by the breakdown of micro-silver halide crystals (chlorides, bromides, iodides), responsive to UV rays and contained in the glass mixture.

The transparency of these systems is related to the level of light striking the glass surface, as they adjust their transmission characteristics in response to intensity, duration, and type of incoming solar radiation. The more global solar radiation hits the glass pane, the darker it tints. The specific configuration of the system's response to varying global solar radiation levels can differ based on manufacturing methods, and it can be tailored to align with the preferences and requirements of users (**Figure 6**).

Generally, SHGC of PC systems varies between 0.48 and 0.31 (for the clearest state) and 0.41 and 0.22 (for the darkest state), while VLT varies between 0.78 and 0.13 (for the clearest state) and 0.73 and 0.09 (for the darkest state) [58].

Due to the ambient temperature's influence on the coloring process of photochromic glasses, with more pronounced effects at lower temperatures and minimal effects at higher temperatures, their applicability in building contexts is significantly limited [59].

Other relevant technologies are PCM systems, which incorporate Phase Change Materials to manage and reduce the energy demands of buildings during peak hours and mitigate fluctuations in building temperatures[3]. These systems leverage the concept of latent heat thermal storage (LHTS) to absorb energy and subsequently release it at different times.

Variable substances are viable to be used as PCMs, enabling the regulation of temperatures within a particular range, determined by the chosen material. PCM glasses typically enable the soft dispersion of natural light within spaces: In their solid state, PCMs allow around 28% of visible light to pass through, whereas in their liquid state, VLT rises to more than 40% [61]. Despite the existence of translucent PCMs [62], the effectiveness of these technologies in terms of transmitting high-quality light remains a drawback. Furthermore, PCMs still have limitations, including challenges associated with selecting the appropriate melting temperature, concerning about the

Figure 6.
Transition phase in a thermochromic glazed façade.

[3] In a study [60], it was found that wall and indoor air temperature fluctuation is decreased by 2.7°C and 1.4°C, respectively, in a building that incorporates PCM; moreover, also energy demand was reduced by 57% during winter.

flammability of paraffin, and the complexity of efficiently dissipating thermal energy from the material following extended periods of elevated temperature exposure.

3.3 Renewable solar energy

By considering that around 40% of the worldwide energy demand is consumed by buildings, "solar Architecture is not about fashion, but about survival," as Architect Norman Foster said, becomes a reality. Anyway, the planning of buildings with multifunctional, integrated façade elements capable of fulfilling the technical demands becomes an essential part of the architectonic mainstream and can contribute to an esthetic valorization.

3.3.1 Semi-transparent PV

The goal of energy change linked to greater use of renewables is successfully achieved when visual appeal and energy efficiency merge together: This architectonic feature finds its optimum in the semi-transparent photovoltaic systems. At the basis of the photovoltaic panel functionality, there is its ability to absorb solar energy and convert it into electricity, transforming photons into electrons (**Figure 7**).

The phenomenon is very similar to selective glass: In the case of transparent photovoltaic panels, a transparent luminescent solar concentrator (TLSC) is used to make the panels completely transparent like glass and, on the other hand, to allow them to absorb wavelengths of light nonvisible to the human eye, such as infrared and ultraviolet light [63].

Today, Building Integrated Photovoltaic (BIPV) can provide optimum U-value (ranging from 0.5 in triple glass glazing to 1.1 W/m^2K in double glass), with optimum

Figure 7.
The Hauptbahnhof, Berlin (Germany): Detail of the 1700 m^2 curved surface covered with 780 semi-transparent c-Si panels (Energy output: 180 kWp, Architect: Meinhard von Gerkan; System provider: Optisol, 2003).

solar factor (G value) and light transmission (TL value) and in the main time to produce solar energy.

Semitransparent PV is formed by Solar PV Cells placed between two panels of glass.

The light transmission and the level of shading inside the building can be controlled and regulated by adjusting the distance between solar PV cells. The panels become transparent when solar PV Cells are positioned far apart; instead, when the cells are positioned closely together, they become semi-transparent and produce a shading effect. Apart from generating electricity, modules can be customized in different dimension, thickness, shape, and color.

Efficiency is quite lower comparing to traditional polycrystalline PV panels.

The average efficiency in intermediate seasons reaches values of about 7.5%, while the conversion efficiency is always calculated as the ratio between generated power and incident radiation, it reaches values equal to 15.5% [64].

Anyway, as this technology is widespread and can be used on a large scale, for example, to cover entire façades of buildings, its lower efficiency is destined to be overcompensated with greater surface development [65].

Semitransparent PV can reduce the carbon footprint of a building, improve thermal insulation, acoustic insulation, and comfort increase, and in general increase the environmental and sustainable value of a building, being a solution that joins functionality, utility, and design.

3.3.2 Bio-adaptive glasses

A novel emerging category of energy-generating glazed systems are bio-adaptive glasses, realized employing photobioreactors, commonly featuring algae as their main component.

Photobioreactors are transparent structures housing a "culture medium" that contains nutrients (typically water), in which microalgae circulate in accordance with the intensity of direct sunlight exposure [66]. These microalgae are consistently supplied with nutrients, and sunlight facilitates photosynthesis, leading to a responsive adaptation of solar shading levels.

Microalgae are often favored over other plant varieties due to their remarkable growth rates and their ability to sequester more CO_2 as they can even double their volume within a week. Moreover, their capability to grow vertically makes them an ideal choice for integration of photobioreactors into building components.

The biomasses produced by bioreactors (the algae) can subsequently be collected for energy-generating purposes (i.e., as biogas to heat water); at the same time, they can capture solar-thermal heat, providing an energy source used to power the building.

This technology has been installed in buildings for the first time in the BIQ house, during the International Building Exhibition in Hamburg in 2013. SolarLeaf's bioreactors, conceived and developed by Arup in cooperation with SSC (Strategic Science Consult of Germany), have four glass layers. The two inner panes are designed with a cavity capacity of 24 liters to facilitate the circulation of the growth medium. Insulating argon-filled gaps flank these panes, contributing to the reduction of heat loss. The front glass panel is composed of white anti-reflective glass, while the rear glass panel offers the option for incorporating decorative glass treatments. A total of 129 modules of photobioreactors, each measuring 70 cm in width, 270 cm in height, and 8 cm in thickness, have been integrated into the façade of this building. With a cultivation area spanning 200 m^2, this system produces 900 kg of biomass annually and generates

Figure 8.
SolarLeaf bioreactor installed in the BIQ House in Hamburg (2013, Arup and Splitterwerk Architects).

around 6000 kWh/year of energy. This energy output is sufficient to cater to the heating requirements of 4 units within the building [67]. For comparison, photovoltaic systems have an efficiency of 12–15% and solar thermal systems of 60–65% (**Figure 8**).

3.3.3 Emerging technologies

Strategies to improve glazing performance can be grouped into four family approaches: The first is the most "traditional one," which acts on the interspace of multilayer glazing assembly (IGU) by using films, low-conductance gases of thermally improved edge spacers, to improve insulation capacity of the system; the second is aimed at altering material composition (e.g., tinted glazing); the third approach involves the application of coatings onto the glass surface to alter hoe it reflects light (such as selective, reflective of low-e coatings). Finally, the fourth resorts to external inputs (whether they are passive or active) to modify the optical characteristics of the glass.

Despite their validity, these approaches still have some limitations; to overcome such issues, researches aimed at discovering new emerging solutions are existing. Some investigate the development of self-regulating window materials as an alternative to glass, such as the reversible thermochromic transparent bamboo smart windows [68], prepared by impregnating delignified bamboo (DB) with epoxy resin containing thermochromic microcapsule powders (TMP), which is colorless at high temperatures and purple at low temperatures, or the air-sandwich glazing systems [69], based on the idea of a set of plastic films, with spacers and air trapped in-between, used as insulation.

Other promising technologies are those resorting to energy storage strategies. One such example is the High Thermal Energy Storage Thermoresponsive (HTEST) smart window [70], which encloses a hydrogel-derived liquid within glass panels. These panels exhibit impressive thermoresponsive optical attributes, boasting 90% luminous transmittance and 68.1% solar modulation. Additionally, the hydrogel-based liquid possesses remarkable specific heat capacity, contributing to the exceptional energy conservation performance of the HTEST smart window.

Unconventional types of smart windows also exist, including humidity-triggered smart windows, that alter light transmittance or window color based on humidity variations. This adaptation influences the transmission of both luminous and near-infrared (IR) light, leading to modifications in transparency or coloration. There are also mechanochromic smart windows, constructed from optically responsive materials that undergo reversible structural adjustments, such as modifications in surface morphology and configuration, in reaction to basic mechanical strain. Consequently, these changes influence optical transmittance through scattering or diffraction of visible light. Moreover, magnetochromic smart windows operate by responding to magnetic field intensity. Changes in magnetic field strength cause nanoparticles to move closer together or farther apart, thereby controlling the smart window's behavior [71].

Again, some concepts concerning emerging glazing technologies are present in literature even if, to the authors' knowledge, they still do not find real development of market applications. Some of them are the Vacuum Tube Window Technology [72], described as a combination of evacuated glass tubes and a glazed frame with Argon in the air gap between them, the Water-flow window [39], originated from the concept of removing the heat stored inside IGU's air-gap thanks to water flooding; the Solar-pond window [69] aimed at integrating into fenestration functions of lighting, heat collection, heat storage, heat preservation and photoperiod control, and the self-sufficient smart window [73] able to regulate the amount of light entering the buildings, varying its color from a transparent state to a blue state without adding energy electricity.

4. Conclusions

This study has allowed defining a synthetic but comprehensive set of requirements concerning several complementary aspects of glazed building components, ranging from architectural to technological, energetic, and economic features of innovative products for advanced glazed skins.

The relationship of façade systems with other building construction technologies has led to emphasizing and underlining the general opportunities for smart materials and solutions in the construction process as a technological and architectural chance, in a sustainable and affordable way.

To summarize the most relevant factors for each category of intervention, a multi-criteria matrix has been defined, in which the façade solutions are briefly described and compared qualitatively by crossing the strategies of interventions with the main requirement set in **Table 1**, Paragraph 3.

The comparison between the strategies identified can be found in the tables included in Appendix A.

This can become a useful tool to define, at first glance, and design criteria and operational tools to guide the design of innovative envelopes, allowing targeted choices to be made about the foreseen interventions, to obtain the desirable levels of quality.

However, it has to be said that despite their potential to reduce energy consumption in buildings while increasing user's comfort[4], the general application of smart

[4] Compared with traditional static windows, smart windows reduce total building energy consumption by approximately 10% [71].

glazed façades is currently hampered by economic and technological factors. Although some of the technologies presented are already in the market, their widespread application is limited because of their elevated price and relatively low fatigue resistance; the main driver for building owners to install smart windows is the desire to eliminate the need for attached shades to allow full access to the outdoor views [74] even if very limited analyses are available on assessing their energy efficiency potential when deployed for residential buildings.

Besides this general overview, the real suitability of each component must be specifically evaluated for each single requirement in each context depending on the design needs.

Acknowledgements

The chapter is the results of a common reflection of the authors, based on their expertise in the specific fields of research.

Specifically, Silvia Brunoro was responsible of the Introduction "The evolution of glass façades performances. State of the art and scientific background" and of the sections "3. Methodology. Categories of requirements and performances," "3.1 Passive solar heat gain: double layer glass façades," "3.3. Renewable solar energy," and "3.3.1. Semi-transparent PV."

Valentina Frighi writes the whole section "2. The role of glazed components," and sections: "3.2 Strategies to control summer overheating," "3.3.2. Bio-adaptive glasses," and "3.3.3. Emerging technologies."

Conclusions are attributable to both the authors, so as the final revision of the work and the approval of the manuscript version to be published.

This work has been funded thanks to the FIRD2022 project "AWARE A multiscalar approach for a resilient and adaptive built environment" grant-aided by the Department of Architecture of the University of Ferrara.

Conflict of interest

The authors declare no conflict of interest.

Appendix

See **Tables A1–A3.**

Glass building systems for façades	
	STRATEGY 1 – Passive solar gain
Criteria for glass façades building components	Double layer glass façades
Architecture	
Transparent	X
Semitransparent	—

Glass building systems for façades	
	STRATEGY 1 – Passive solar gain
Independent structure	X
Integrate in façade	—
Architectural and smart media properties (e.g., multimedia)	—
Energy properties and comfort	
Visual Lighting transmission VLT (%)	50–70
Solar factor SHGC (%)	10–40
Thermal insulation U value (W/m^2K)	0.8–1.5*
Greenhouse effect (solar gain)	X
Thermal inertia	X
Acoustic insulation (Db)	3–10
Natural ventilation	X
Solar devices integration	X
Renewable energy integration	—
Environmental properties	
Recyclability	—
Green eco label products	X
Pollution adsorption (e.g., photocatalytic)	—
Economic features	—
Construction cost (€/m²)	1200–2000

**The "thermal insulation" values (*U-value*) are referred to a technical situation of a single IGU (6 mm + 12 mm + 4 mm) Air-filled.*

Table A1.
Matrix of the main features of façade solution systems for comparison: STRATEGY 1 – Passive solar gain.

Glass building systems for façades													
	STRATEGY 2 – Summer overheating control												
Criteria for glass façades building components	TINT.	REFL.	LOW-E	SEL.	VIG	TIM	HG	EC	GC	LCD	TC	PC	PCM
Architecture													
Transparent	X	one side view	X	X	X	—	X	X	—	X	X	X	X
Semitransparent	—	—	—	—	—	X	—	—	X	X	—	—	X
Independent structure	—	—	—	—	—	—	—	—	—	—	—	—	—
Integrate in façade	X	X	X	X	X	X	X	X	X	X	X	X	X
Architectural and smart media properties (e.g., multimedia)	—	—	—	—	—	—	X	X	—	X	—	—	—
Energy properties and comfort													
Visual Lighting transmission VLT (%)	30–70	30	70–80	60	78	68	70–75	0/60	10–60	50/70	8/55	6/60	8–28/12–44
Solar factor SHGC (%)	45–70	45–55	50–60	40–45	67	64	50	9/40	10–50	0/60	15/35	10/35	33/4–37
Thermal insulation U value (W/m^2K)	2.8	2.8	1.6	2.7	0.4	0.80–1.3	1.4–2.8	1.80	0.9	1.50	1.36	1.20	0.48
Greenhouse effect (solar gain)	X	X	X	X	X	—	X	X	—	—	—	—	—
Thermal inertia	—	—	—	—	—	X	—	—	—	—	—	—	X
Acoustic insulation (Db)	—	—	—	—	36	—	—	35	—	—	—	—	—
Natural ventilation	—	X	—	—	—	—	—	—	—	—	—	—	—
Solar devices integration	X	—	X	X	X	—	X	X	—	—	—	—	—
Renewable energy integration	—	—	—	—	—	—	—	—	—	—	—	—	—
Environmental properties													
Recyclability	X	X	X	X	N.A.	N.A.	X	N.A.	N.A.	—	N.A.	N.A.	N.A.
Green eco label products	X	X	X	X	N.A.	N.A.	X	N.A.	N.A.	—	N.A.	N.A.	N.A.
Pollution adsorption (e.g., photocatalytic)	—	—	—	—	N.A.	N.A.	—	N.A.	N.A.	—	N.A.	N.A.	—
Economic features													
Construction cost (€/m^2)	20–100	50–90	30–70	50–130	250–500	N.A.	~150	~100	N.A.	85–130	~100	~100	500–1000

***N.A.** = Not applicable, thus meaning considered non-relevant for the state of development of the technologies explained (i.e., technologies still at a prototypal state or not on-the-market available).*
All the "thermal insulation" values (U-value) are referred to a technical situation of a single IGU (6 mm + 12 mm + 4 mm) Air-filled, except for VIG (Vacuum Insulated Glass), considered as "monolithic", and for GC (Gasochromic), for which available data are related to a double IGU gas-filled.
All the values related to VLT and SHGC of chromogenic windows technologies are referred to the two states of the systems (opaque/transparent); for PCM glasses, the values are referred to the two phases of the PCM: from left to right, first numbers reflect the situation when the PCM is crystalized while second numbers refer to PCM at liquid state.
Construction cost is approximate and related to single glass pane, thus means non-referred to a IGU technical solution.

Table A2.
Matrix of the main features of façade solution systems for comparison: STRATEGY 2 – Summer overheating control.

Glass building systems for façades												
	STRATEGY 3 – Renewable solar energy											
Criteria for glass façades building components	ST PV	BAG	RTTB	ASGS	HTEST	HTSW	MCSW(a)	MCSW(b)	VTWT	WFW	SPW	SSSW
Architecture												
Transparent	—	—	—	—	X	—	X	X	—	—	X	X
Semitransparent	X	X	X	X	—	X	X	X	X	X	—	—
Independent structure	—	X	X	X	X	X	X	—	X	X	X	—
Integrate in façade	X	—	—	—	—	—	—	X	—	—	—	X
Architectural and smart media properties (e.g., multimedia)	—	X	X	—	X	X	X	X	—	—	X	X
Energy properties and comfort												
Visual Lighting transmission VLT (%)	30–50	8–50	60–80	50–60	~ 90	6/65	9/92	—	—	0.24–0.67	—	—
Solar factor SHGC (%)	40–60	10–80	—	—	68	—	—	—	—	—	—	—
Thermal insulation U value (W/m^{2}K)	—	~5	—	1.80–3.40	—	—	—	—	0.30–2	0.40	0.40–2	—
Greenhouse effect (solar gain)	—	—	X	—	X	X	X	X	—	X	X	X
Thermal inertia	—	X	—	X	X	—	—	—	X	—	X	—
Acoustic insulation (Db)	—	X	—	X	X	—	—	—	X	X	—	—
Natural ventilation	—	—	—	—	—	—	—	—	—	—	—	—
Solar devices integration	X	—	X	—	—	—	—	—	—	—	—	—
Renewable energy integration	X	X	—	—	X	—	—	X	—	—	—	X
Environmental properties												
Recyclability	X	X	—	—	—	—	—	—	—	—	—	—
Green eco label products	X	—	—	—	—	—	—	—	—	—	—	—
Pollution adsorption (e.g., photocatalytic)	—	X	—	—	—	—	—	—	—	—	—	—
Economic features												
Construction cost (€/m^2)	176–325 €/module	—	—	—	—	—	—	—	130	—	120	—

ST PV *= Semitransparent photovoltaic;* ***BAG*** *= Bio Adaptive Glass;* ***RTTB*** *= Reversible Thermochromic Transparent Bamboo;* ***ASGS*** *= Air-Sandwich Glazing System;* ***HTEST*** *= High Thermal Energy Storage Thermoresponsive smart window;* ***HTSW*** *= Humidity Triggered Smart Window;* ***MCSW(a)*** *= Mechanochomic Smart Window;* ***MCSW(b)*** *= Magnetochomic Smart Window;* ***VTWT*** *= Vacuum Tube Window Technology;* ***WFW*** *= Water-flow window;* ***SPW*** *= Solar Pond Window;* ***SSSW*** *= Self-Sufficient Smart Window.*

Given the extremely innovative nature of the technologies presented in this table, it was not possible to find some data. Therefore, the cells filled with "-" are to be considered empty due to the difficulty or impossibility of finding this information.

Table A3.
Matrix of the main features of façade solution systems for comparison: STRATEGY 3 – Renewable solar energy.

DOI: http://dx.doi.org/10.5772/intechopen.113127

Author details

Silvia Brunoro* and Valentina Frighi
Department of Architecture, University of Ferrara, Italy

*Address all correspondence to: silvia.brunoro@unife.it

References

[1] Banham R. The Architecture of the Well – Tempered Environment. London: Architectural Press; 1969

[2] Joddio P. Building a New Millennium. Bridlington, UK: Taschen; 2000

[3] Jones DL, Hudson J. Architecture and the Environment. Bioclimatic Building Design. London: Laurence King Publishing; 1998

[4] Campioli A. Il contesto del progetto. Il costruire contemporaneo tra sperimentalismo high-tech e diffusione delle tecnologie industriali. Milano: FrancoAngeli; 1993

[5] Wigginton M, Harris J. Intelligent Skins. London: Architectural Press; 2002

[6] Diprose PR, Robertson G. Towards a fourth skin? Sustainability and double-envelope buildings. Renewable Energy. 1996;**8**:169-172

[7] Markus TA, Morris EN. Buildings, Climate and Energy. London: Pitman; 1980

[8] Gonzalo R, Habermann KJ. Energy Efficient Architecture. Basics for Planning and Construction. Basel – Boston – Berlin: Birkhauser; 2006

[9] Hausladen G., De Saldanha M., Liedl P., ClimateSkin: Building-skin Concepts that Can Do More with Less Energy. 2008

[10] Kyoto Protocol To The United Nations Framework Convention On Climate Change. 1997. Available from: http://unfccc.int/resource/docs/convkp/kpeng.pdf

[11] European Commission, European Green Deal, Climate Action. Available from: https://climate.ec.europa.eu/eu-action/european-green-deal_en [Accessed: June 29, 2023]

[12] EEAP. Italian Energy Efficiency Action Plan. Vol. 20172017. Available from: https://energy.ec.europa.eu/system/files/2018-03/it_neeap_2017_en_0.pdf [Accessed: June 26, 2023]

[13] European Commission. Directive 2010/31/EU of 19 May 2010 on the Energy Performance of Buildings (Recast)

[14] Compagno A. Intelligent Glass Facades; Material, Practice, Design. Basel (CH): Birkhauser; 1995

[15] Hall P. Living Skins: Architecture as Interface. 2006. Available from: http://power.nilut.com/documents/Living_Skins.pdf

[16] Colafranceschi D. Sull'involucro in architettura: Herzog, Nouvel, Perrault, Piano, Prix, Suzuki, Venturi, Winesi. Rome: Librerie Dedalo; 1996

[17] Lei Q et al. Active-passive dual-control smart window with thermochromic synergistic fluidic glass for building energy efficiency. Building and Environment. 2022;**222**:109407

[18] Jelle BP et al. Fenestration of today and tomorrow: A state-of-the-art review and future opportunities. Solar Energy Materials & Solar Cells. 2012;**96**:1-28

[19] Al-Obaidi KM et al. A review of skylight glazing materials in architectural designs for a better indoor environment. Modern Applied Science. 2014;**8**:68-73

[20] Bedon C et al. Structural characterisation of adaptive facades in

Europe – Part I: Insight on classification rules, performance metrics and design methods. Journal of Building Engineering. 2019;**25**:100721

[21] Bedon C et al. Structural characterisation of adaptive facades in Europe – Part II: Validity of conventional experimental testing methods and key issues. 2019;**25**:100797

[22] Favoino F et al. Building Performance Simulation and Characterisation of Adaptive Facades. Delft: TU Delft Open; 2018

[23] Morse DL, Evenson JW. Welcome to the glass age. International Journal of Applied Glass Science. 2016;**7**(4): 409-412

[24] Olgyay V. Design with Climate. Bioclimatic Approach to Architectural Regionalism: Bioclimatic Approach to Architectural Regionalism. Princeton University Press; 1980

[25] Oesterle E, Lieb R, Lutz M, Heusler W. Double Skin Facades. Prestel: Integrated Planning; 2001

[26] Ruiz-Pardo A. Simulation model for some types of double envelope elements. The International Journal of Ventilation. 2010;**9**(3):227-239

[27] Lemma HN. Double versus single skin facades in hot arid areas. Energy and Buildings. 2008;**40**:240-248

[28] O'Connor J, Lee E, Rubinstein F, Selkowitz S. Tips for Daylighting with Windows; the Integrated Approach. Berkeley, CA: Ernest Orlando Lawrence Berkeley National Laboratory

[29] Brunoro S. Double-skin ventilated glass façades. An overview of concepts, product, systems, realized examples. In: Memari AM, Lowe SK, editors. 2020 Residential Building Design & Construction Conference Proceedings. Pennsylvania Housing Research Center (PHRC) at The Pennsylvania State University; Sep 2020. pp. 619-631. ISBN: 978-1-62307-007-6

[30] Krewinkel H. Glass Buildings. Basel, Berlin, Boston: Birkhauser; 1998

[31] Stec WJ, Van Paassen AHC, Maziarz A. Modelling the double skin façade with plants. Energy and Buildings. 2005;**37**:419-427

[32] Irulegi O, Serra A, Hernández R, Ruiz-Pardo Á. Ventilated active façades to reduce the cooling demand of office buildings - The case of Spain. International Journal of Ventilation. 2011;**1**:101-113

[33] Loncour K, Deneyer A, Blasco M, Flamant G, Wouters P. Ventilated Double Façade. Classification & Illustration of Façade Concepts. Belgian Building Research Institute, Department of Building Phisics, Indoor Climate & Buidling Services; 2004

[34] Ding W, Hasemi YYT. Natural ventilation performance of a double-skin façade with a solar chimney. Energy and Buildings. 2005;**37**:411-418

[35] Lemma HJ, Bartak M, Drkal F. Modeling and simulation of a double-skin façade system. ASHRAE Transactions. 2002;**108**:1251-1259

[36] Straube J. Analysis of Double Facades for Cool Humid Climates (thesis). Waterloo: Department of Civil Engineering, University of Waterloo; 2001

[37] Pikas E, Thalfeldt M, Kurnitski J. Cost optimal and nearly zero energy building solutions for office buildings. Energy and Building. 2014;**74**:30-42

[38] Allen K et al. Smart windows – Dynamic control of building energy performance. Energy and Buildings. 2017;**139**:535-546

[39] Chow T, Li C, Lin Z. Innovative solar windows for cooling-demand climate. Solar Energy Materials & Solar Cells. 2010;**94**:212-220

[40] Liu H. The Development of Novel Window Systems Towards Low Carbon Buildings. Nottingham: The University of Nottingham; 2012

[41] Cuce E. Toward multi-functional PV glazing technologies in low/zero carbon buildings: Heat insulation solar glass – Latest developments and future prospects. Renewable and Sustainable Energy Reviews. 2016;**60**:1286-1301

[42] Sun Y, Wu Y, Wilson R, Sun S. Thermal evaluation of a double-glazing façade system with integrated Parallel Slat Transparent Insulation Material (PS-TIM). Building and Environment. 2016; **105**:69-81

[43] Qu J et al. Transparent thermal insulation coatings for energy efficient glass windows and curtain walls. Energy and Buildings. 2014;**77**:1-10

[44] Baetens R, Jelle BP, Gustavsen A. Aerogel insulation for building applications: A state-of-the-art review. Energy and Buildings. 2011;**43**: 761-769

[45] Sun Y, Wilson R, Wu R. A review of transparent insulation material (TIM) for building energy saving and daylight comfort. Applied Energy. 2018;**226**: 713-729

[46] Gorgolis G, Karamanis D. Solar energy materials for glazing technologies. Solar Energy Materials & Solar Cells. 2016;**144**:559-578

[47] Tavares PF et al. Evaluation of electrochromic windows impact in the energy performance of buildings in Mediterranean climates. Energy Policy. 2014;**67**:68-81

[48] Cannavale A et al. Perovskite photovoltachromic cells for building integration. Energy & Environmental Science. 2015;**8**:1578-1584

[49] Bogati S, Georg A, Graf W. Photoelectrochromic devices based on sputtered WO and TiO films. Energy Materials & Solar Cells. 2017;**163**:170-177

[50] Feng W et al. Gasochromic smart window: Optical and thermal properties, energy simulation and feasibility analysis. Solar Energy Materials & Solar Cells. 2016;**144**:316-323

[51] Rezaei SD, Shannigrahi S, Ramakrishna S. A review of conventional, advanced, and smart glazing technologies and materials for improving indoor environment. Solar Energy Materials & Solar Cells. 2017;**159**: 26-51

[52] Mesloub A et al. Assessment of the overall energy performance of an SPD smart window in a hot desert climate. Energy. 2022;**252**:124073

[53] Kim Y et al. Optical properties and optimized conditions for polymer dispersed liquid crystal containing UV curable polymer and nematic liquid crystal. Current Applied Physics. 2015; **15**:292-297

[54] Hee WJ et al. The role of window glazing on daylighting and energy saving in buildings. Renewable and Sustainable Energy Reviews. 2015;**42**:323-343

[55] Li SY, Niklasson GA, Granqvist CG. Thermochromic fenestration with

VO2-based materials: Three challenges and how they can be met. Thin Solid Films. 2012;**520**:3823-3828

[56] Aburas M et al. Thermochromic smart window technologies for building application: A review. Applied Energy. 2019;**255**:113522

[57] Casini MS, Buildings. Advanced Materials and Nanotechnology to Improve Energy-Efficiency and Environmental Performance. Oxford (UK): Elsevier; 2016

[58] Tällberg R, Jelle BP, Loonen R, Gao T, Hamdy M. Comparison of the energy saving potential of adaptive and controllable smart windows: A state-of-the-art review and simulation studies of thermochromic, photochromic and electrochromic technologies. Solar Energy Materials and Solar Cells. 2019;**200**:109828

[59] Granqvist GC et al. Progress in chromogenics: New results for electrochromic and thermochromic materials and devices. Solar Energy Materials and Solar Cells. 2009;**93**: 2032-2039

[60] Nghana B, Tariku F. Phase Change Material's impacts on the energy performance and thermal comfort of buildings in a mild climate. Building and Environment. 2016;**99**:221-238

[61] Goia F. Thermo-physical behaviour and energy performance assessment of PCM glazing system configurations: A numerical analysis. Frontiers of Architectural Research. 2012;**1**:341-347

[62] Long L et al. Performance demonstration and evaluation of the synergetic application of vanadium dioxide glazing and phase change material in passive buildings. Applied Energy. 2014;**136**:89-97

[63] Polo López CS, Sangiorgi M. Comparison assessment of BIPV façade semi-transparent modules: Further insights on human comfort conditions. Energy Procedia Journal. 2014;**48**: 1419-1428

[64] Bortolini M, Gamberi M, Graziani A, Mora A, Regattieri C. Multi-parameter analysis for the technical and economic assessment of a photovoltaic systems in the main European Union countries. Energy Conversion and Management. 2013;**74**:117-128

[65] Polo LC, Frontini F, Bonomo P, Scognamiglio A. Pv And Façade Systems for the building skin. Analysis of design effectiveness and technological features. In: 29th European Photovoltaic Solar Energy Conference And Exhibition

[66] Villalba MR, Cervera R, Sánchez J. Green solutions for urban sustainability: photobioreactors for algae cultivation on façades and artificial trees. Buildings. 2023;**13**:1541

[67] Vajdi S, Aslani A. Design and techno-economic analysis of direct CO2 capturing with integrated photobioreactors as a building façade. Sustainable Energy Technologies and Assessments. 2023;**56**:103068

[68] Ji Y et al. Reversible thermochromic transparent bamboo for dynamically adaptive smart windows. Industrial Crops and Products. 2023;**197**: 116593

[69] Cuce E, Riffat SB. A state-of-the-art review on innovative glazing technologies. Renewable and Sustainable Energy Reviews. 2015;**41**:695-714

[70] Zhou Y et al. Liquid thermo-responsive smart window derived from hydrogel. Joule. 2020;**4**:2458-2474

[71] Zhou Y et al. Unconventional smart windows: Materials, structures and designs. Nano Energy. 2021;**90**:106613

[72] Cuce E, Riffat SB. Vacuum tube window technology for highly insulating building fabric: An experimental and numerical investigation. Vacuum. 2015; **111**:83-91

[73] Wang J et al. A bi-functional device for self-powered electrochromic window and self-rechargeable transparent battery applications. Nature Communications. 2014;**5**:4921

[74] DOE. Emerging Technologies Research and Development, Research and Development Opportunities Report for Windows, Report for Building Technology Office. Washington, D.C: US Department of Energy; 2020. Available from: https://www.energy.gov/sites/prod/files/2020/05/f74/bto-20200504_Draft_Windows_RDO.pdf

Chapter 7

The 7D BIM Model Used in the Estimation of the Useful Life of Façade Materials

Alcínia Zita Sampaio, Inês Domingos and Augusto Gomes

Abstract

The Building Information Modelling (BIM) methodology is supported on the concept of centralizing, in a parametric virtual model, all information related with the project, the construction and the overall lifecycle of a building. The building maintenance and management activities, requires the development of working facilities planned in an early project phase. The maintenance planning has been improved supported in BIM, as it allows professionals to easily retrieve, add and update the database of the BIM model. The definition of adequate maintenance strategies requires knowledge regarding the durability of the materials, mainly the degradation perdition of the materials. The present work is focused on the estimation of the useful lifetime of materials usually applied on the finishing of traditional building façades and terraces. Based on the knowledge of durability of the selected materials a Dynamo script was created allowing to obtain an estimation value concerning the degradation perdition of the materials. Other Dynamo script was developed oriented to the visualization of the degradation level of the materials. This innovative approach intends to support the maintenance engineers to make assertive decisions concerning the maintenance activity. In this study Dynamo programming improved BIM-FM systems integration, providing a positive contribution in construction maintenance context.

Keywords: BIM, 7D model, management planning, degradation prediction, wall's finishing, dynamo script

1. Introduction

Building Information Modelling (BIM) methodology offers a rigorous and complete representation of the building and presents a great potential to support the design, construction and occupation phases. However, the implementation of BIM requires an initial effort to train human resources, in order to acquire the skills mandatory in the use of the available tools, and a financial investment in the acquisition of new technologies and equipment.

The construction industry has been progressively incorporating the implementation of BIM methodology, supported by the permanent dissemination of its benefits recognised in reports and in recent technologic advances [1]. This methodology reveals a new and different way of working in the construction sector when compared to traditional methods based on two-dimensional (2D) technical documents. The BIM work transmits more transparency, better communication and a greater level of collaboration between all partners involved in the development of building projects [2]. When developing a project, using BIM platforms, a digital three-dimensional (3D) model of the physical and functional characteristics of the building in analyses is created. The model brings together the information produced in the various phases of the building life cycle forming a complete database supporting decision-makers [3].

The Facility Management (FM) activity, in a building, covers several disciplines in ensuring the functionality of the built environment integrating people, places, processes and technologies. The principal advantage that BIM can bring to this discipline is mainly related to the great capacity of storing a large volume of information, which normally is required in the maintenance activity [4]. The capacity to archive a great volume of data and allowing its accessibility with the connection to several FM systems make BIM strategies a very interesting perspective. During the occupation of a building, BIM implementation within the management and maintenance phases is not yet a recurrent concept, despite the benefits that have been reported. The research in this area, although it is currently growing, is still at an early stage [5].

The maintenance of a building starts immediately after its construction and ends with its demolition. It creates an accurate depiction of the physical conditions, environment, and assets of a facility. Currently, BIM-FM integration is still unusual. However, retrieving data from the model requires experience and relevant waste of time [6]. The FM professional is the 'expert in charge of a building whose concern covers operational issues of maintenance, cleanliness and safety of tenants'. The growing complexity of buildings and the significant cost of their administration has led to the need to introduce 'strategic and tactically based management functions', associated with other support activities such as the control of the human resources involved and the use of adequate and integrated software.

The present study intends to contribute to the reuse the information created and stored in the database of the model and its application in the maintenance activity in an integrated way [7]. The text introduces the current practices in BIM-FM and the main advantages and limitations in using BIM platforms, in the maintenance context. The study considers the development of two Dynamo scripts created to improve the integration of BIM models to FM functionalities. This innovative approach intends to support the management professionals to define adequate maintenance planning with the use of the created Dynamo scripts, to estimate service life of some frequent materials applied as finishing of exterior walls and roofs.

2. BIM-FM integration

The concept of Facility Management (FM) is understood as an interdisciplinary practice related to the management of buildings and facilities organised by the experts involved. The distinct applicability of BIM is referred to in the literature as the 'n' dimensions of BIM. Each dimension is associated with a set of distinct information, added to the 3D model, complementing it with the activities that can be elaborated, based on a selective retrieving and extraction from the 3D BIM model database [8]: the time factor (4D); the quantity and costs estimation (5D), simulation of energetic consume (6D) and the operation process (7D). So, the seventh dimension comprises the application of BIM methodology in the Facility Management activity. This activity supported in BIM platforms can be improved with recognised benefits. It can enable procedures like archiving, retrieving and reusing a great volume of data that is normally required in the context of the activity [9].

The 7D model gathers the information regarding each component and space, the applied materials, their quantification and costs and the specific information required to support maintenance actions. These data englobes data relate to material enterprises, to technical documents the products, to warranties and to guide manuals [10]. The 7D model supports the professional to take on the management phase of every achieved and retrieved data. By combining the quality of the facilities with the active control of costs, the FM allows to increase the efficiency of the organisation that uses the building and improves the quality of the employees' work [11].

The FM concept is still relatively recent in the construction industry. The International Facility Management Association (IFMA) was created in the United States of America in 1978 and was integrated in 1984 into the European entity by the European Facility Management Network (EuroFM), in order to implement FM in the practice of construction, education and research. Since then, its importance has been increasing and recognised as necessary in the management of a building, along with its occupation [12]. Based on the usual requirements and practices used in FM, IFMA has identified the 11 main competencies related to FM activity [13]:

- Communication: definition of plans and processes within the internal and external stakeholders;

- Quality: best practices, process improvements, and audits must be applied;

- Occupation and human factor: healthy and safe work environment for users;

- Finance and business: strategic plans, budgets, financial analysis and acquisitions;

- Emergency and business continuity plan: emergency and risk management plans and procedures;

- Leadership and strategy: strategic planning, organisation, team and leaders;

- Project management: supervision and management of all related projects and contracts;

- Environmental management and sustainability: sustainable management of natural and built environments;

- Real estate and property management: planning and acquisition of real estate;

- Operation and maintenance: operation and maintenance of buildings, services to occupants;

- Technology: FM technology, workplace management systems.

As the generation of a BIM model is based on the use of parametric objects, each component of a building includes a large amount of data such as the type of the applied materials and its related proprieties. In BIM-FM integration, the degree of interoperability performed between different software is the main limitation in applying BIM in the operational phase of the building [14]. However, using the native data format of the modelling system, the transfer process of information from BIM model to FM systems can be effectuated without loss of relevant information. Additionally, the model can be permanently updated, allowing the facility manager to update the building's operation services [15]. Therefore, it is possible to contribute to the reduction of investment costs and better facilities management services in a building [16].

Technological advances in BIM software and integration capacity aimed to improve the efficiency in supporting FM practices. BIM-FM integration allowed the development of a more comprehensive and reliable computational solution related to the collection, categorisation, visualisation and updating of information about the operation and maintenance of a building. This information must be complete, organised, accurate and accessible. The main challenge is to integrate the gathered information with the data acquired in the design and construction phases and the storage of this data in the BIM model, using adequate building management and operation systems [17].

The life cycle of a building involves several stages: The design of the project; the construction process; the building occupation; the demolition procedure. The phase of longest duration corresponds to the occupation of the building. To maintain the quality of the building, throughout this phase, it is necessary to carry out maintenance, conservation or renovation actions applied to the components of the building, in order to ensure the service conditions and to prolong the useful life of the building.

3. Estimation of the service life of materials

The service life period considers the definition of a minimum acceptable level for the performance of a building, depending not only on the evaluation criteria of the expert but also on the safety aspects, functionality for a given time or the environmental or regulatory context in which the assessment carried out is inserted. Considering this great diversity of influences, the prediction of the useful life of a building is difficult to estimate with accuracy. The value of the predate service life is estimated based on simple and linear mathematic models. In addition, a building does not age in a homogeneous way and its surroundings are subject to a high number of aggressive agents, resulting in a faster ageing rate when compared to the interior components [18].

The ageing process of the building encompasses a chain of events, directly conditioned, by the decisions implemented in the design phase and by the use or occurrences developed during the occupation period. The ageing mode also depends on the functional changes and constructive changes imposed, resulting from the technical evolution of aesthetic and comfort requirements, according to the users' needs. The components of the building are subjected to a diversity of deterioration processes, contributing to the ageing phenomenon [19]. The following prediction methods can be considered deterministic, probabilistic (or stochastic) and engineering methods [20].

The method considered in the study is the factor deterministic procedure. This method is a simplified approach that allows to calculate the estimated service life

(ESL) value for the construction materials. This value is obtained by the product of the reference service life (RSL) value by several values related with a set of deterministic factors (**Table 1**).

- The RSL value is data delivered by the construction company or homologated documentation, and by the set of data confirmed in analogous construction and behind similar exterior conditions of occupation;
- The factors from A to G can present values between 0.8 and 1.2. The unit value is considered when the factor has no weight for the material, a higher value increases the ESL; a lower value decreases the estimated service life.

In order to obtain a correlative database, the bibliographical references of Raposo [21], Lopes [22], and Matos [23] were checked. In it, three cases of application of the factor method according to the guidelines of the ISO 15686-1 standard were analysed [24]. The preliminary bibliographic research was conducted to define the durability matrix of the database. The retrieved data concerns the required factors and are related to the materials analysed in the present study, namely, cladding applied on the flat roof (**Table 2**), adherent ceramic tiling finishing one façade and the ventilated façade tiling type. In the aim of the study, the developed Dynamo script uses the factor values listed in independent tables concerning each material.

ESL = RSL × factor A × factor B × factor C × factor D × factor E × factor F × factor G		
Factor A—quality of components	Factor B—design level	Factor C—work execution level
Factor D—indoor environment	Factor E—outdoor environment	Factor F—in-use conditions
Factor G—maintenance level		

Table 1.
The factors that interfere with estimation of the service life value.

Factor	Description		Value
Factor A	A1	Declaration of Conformity and Quality Certificate	
		With EC Declaration of Conformity and Quality Certificate	1.2
		With EC declaration of conformity or Quality Certificate	1.0
	A2	Characteristics of thermal insulation of the cover	
		With ISOLE rating higher than recommended for coverage type	1.2
		With the recommended ISOLE rating for the coverage type	1.0
		With ISOLE rating lower than recommended for coverage type	0.8
	A4	Type of bituminous membrane armour	
		Polyester felt of 250 g/m^2 or armed with polyester and fibreglass felt	1.2
		With polyester felt of at least 150 g/m^2	1.0
		With polyester felt less than 150 g/m^2 or with fibreglass felt	0.8

Table 2.
Factor A of the durability matrix of cladding applied on terrace roofs.

4. Visual programming in dynamo

Deterministic methods are based on the analysis of the factors and mechanisms that affect the degradation of the constructive elements, under the normal conditions of use, allowing for quantifying the level of degradation of the material. The present study proposes to include the factor method for estimating the service life of elements in construction. It involves the development of two scripts, using the visual programming tool Dynamo [25]: the 'Estimate Service Life' script and the 'ESL Analysis'. The scripts were applied over a case study allowing verifying its efficiency.

4.1 Dynamo script 'Estimate Service Life'

The tables of durability matrices previously created were used to support the development of the script 'Estimate Service Life'. The tables were integrated into the script so that the options of each factor can be selected according to the element and material used in order to estimate the respective service life value. The script can run over projects modelled in Revit and applied over the considered constructive elements: roof waterproofing membranes; adherent ceramic tiling on façades; ventilated façade claddings.

After obtaining the ESL value, for each of the elements under study, it is possible to make a prediction of the state of degradation of the elements for the horizon of a given year. The degradation process follows an incremental and uniform evolution. A set of colours was assigned according to the range in which the calculated % of useful life is. The script 'Estimate Service Life' was created using Dynamo visual programming whose main structure is represented in **Figure 1**.

In the development of the script 'Estimate Service Life (ESL)', it required the installation of the package Data-Shapes, which allows the programmer to collect

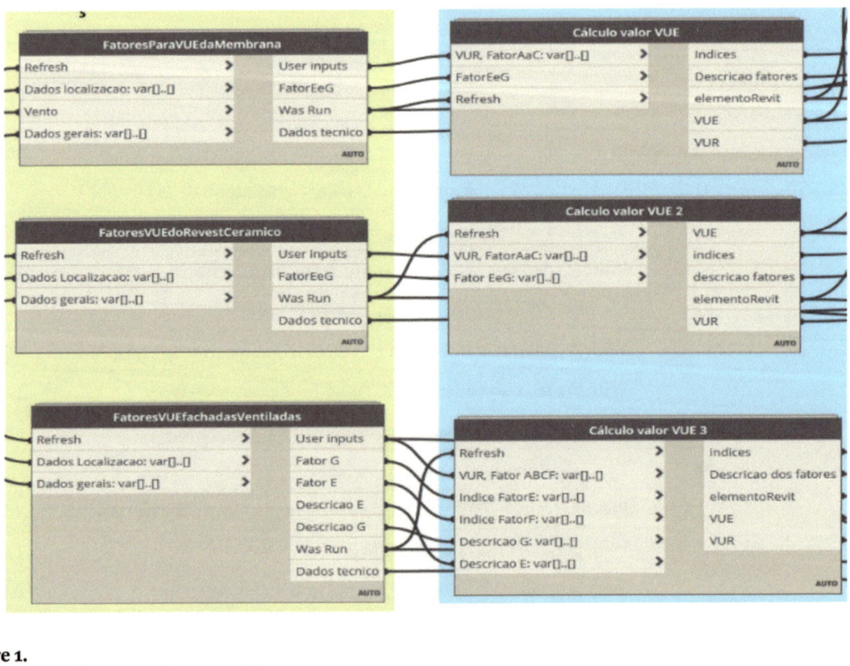

Figure 1.
Dynamo script 'Estimate Service Life'.

various types of user inputs depending on the type of data to be inserted. This package is very flexible, allowing an easy control of the type and order of input data to be considered. The Code Block node, available in Dynamo standard library, is responsible for the insertion of various types of data. The connection to each entry in the master node is made through the use of lists. Due to the high volume of information to be presented in the interfaces, it was decided to use the Python Script node for the creation of the lists, simplifying the programming in the Dynamo interface.

To connect the tables inserted in the database to Dynamo it was necessary to instal the Slingshot package. This package presents a collection of nodes used in the database management system. They allow the connection and communication to the databases created in MySQL and SQLite formats, using the SQL language.

The inputs of the Dynamo script can be easily adapted to the needs of the user. The scripts can be configured to request input data prior to triggering the script. To perform this the Boolean node, selected as an input command, accessed in Dynamo menu, was set as input data for Dynamo Player. The Boolean (True or False) node displays the false option and deletes the data recorded in the last use of the script; the true option allows you to run the program and display the interface for a new use.

The selection of the project site allows us to suggest filling options for the factor E (to be addressed later) related to the characteristics of the outdoor environment. As an example of localization, the Lisbon, Porto and Algarve options were incorporated, chosen because they are associated with the examples used in the script programming. The interface also allows you to enter the type of fixation of the coating, variable according to the support structure chosen for the ventilated façade, and the data for identification of the technician. In this case, the fastening system chosen is a sight fastening system using clips.

4.2 Generation of the 'ESL analysis' script

The 'ESL Analysis' script was developed in association with the 'Estimate Service Life' script and allows:

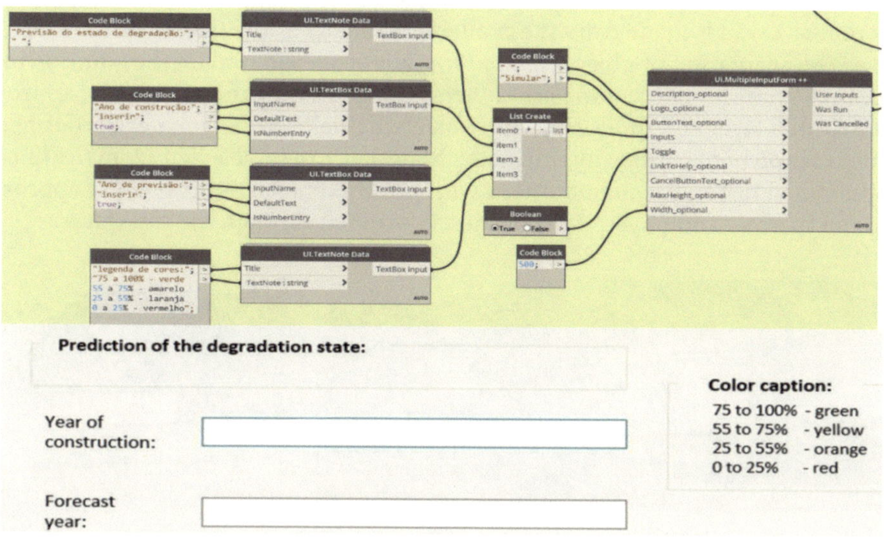

Figure 2.
Dynamo script 'ESL analyses' and the assignment of colours.

- Calculate the % lifetime for a given year, based on the value calculated for the ESL parameter;

- Assign a colour scheme to elements based on the % lifetime.

For each model component associated with the ESL parameter, the obtained ESL value is assigned. Based on it is possible to make an evaluation of the degradation state of the material for a given year. According to the value of ESL, the related percentage is considered, and the component presented the colour assigned to the respective interval of values. An initial degradation state corresponds to the green colour and an advanced degradation state is related to the red colour. In the second script, a set of colours was allocated according to the level of the achieved deterioration, in which the *% of lifespan* is established (**Figure 2**).

5. Application of the scripts

The selected building case, located in Lisbon, is composed of eight floors and a terrace. Using the Revit software (Autodesk), the architectural BIM model was created, having been modelled with a great detail, namely, the coatings applied to the façades and the roof (**Figure 3**).

- The north façade is composed of a double wall in brick masonry and presents a finishing material of porcelain tiles;

- The southeast façade presents a construction solution in ventilated clad with fibre cement plates;

- The roof terrace presents a finishing material composed of a waterproofing membrane in polymer bitumen.

For the representation of the walls, new parametric objects were created. A first selection of a wall is made from the available elements of the library of the system in use. The adaptation of objects related to walls is made based on the addition of successive layers in their composition, being assigned the type of material appropriate to each layer and indicating the respective thickness. The materials defined as membranes do not assume any thickness. Materials can be selected from the library of predefined materials, characterised with parameters related to graphic appearance and thermal and physical properties, controlled through the interfaces.

a *b* *c*

Figure 3.
Main façade with tiles (a), rear façade with fibre cement (b) and flat roof (c).

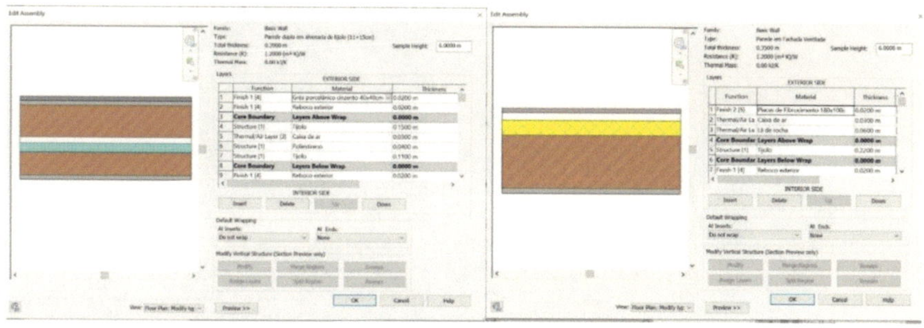

Figure 4.
Double wall in brick masonry (11 + 15 cm) with porcelain stoneware coating and wall in ventilated façade with fibre-cement plates.

The materials available in the system can be adapted to the graphic characteristics and properties required in each specific case. In the present case, there were selected materials close to those required, and later carried out the modelling of the outer cladding layers of the walls, adapted to the intended solutions. The layers of the coating materials applied to the façades were porcelain stoneware and fibre-cement boards (**Figure 4**).

The parametric objects used in the modelling process of the building were incremented with the necessary parameters related to the ESL of the components under analysis. The mentioned parameters, ESL and % lifespan, were assigned to the materials applied in both façades and the terrace roof. This allows the user to calculate the respective values obtained from the generated Dynamo scripts:

- The parameter ESL accepts the value estimated by the 'Estimate Service Life' script;
- The Excel file Maintenance records the location of the generated Excel file by the some script;
- The parameter % lifespan accepted the variable calculated based on the ESL parameter and the forecast year, obtained by the script 'ESL Analysis'.

5.1 Execution of the 'Estimate service life' script

The execution of the script is running through Dynamo Player option included in the upper menu of Revit. The script was successfully executed to estimate the service life of each of the materials identified and the result is visualised in **Figure 5**.

A table with the numerical results can be accessed and the aspect of the deterioration is illustrated by the colour assumed by each element. A period of ten years, from year 2020 and 2030, was tested. These years were referred to as the construction stage and a life forecast year.

5.2 Execution of the 'ESL analysis' script

Like the 'Estimate Service Lifetime' script, the 'ESL Analysis' script can be triggered through the Dynamo Player window. Having been configured as inputs of this script the selection of the elements in the created view and the Boolean node,

Technician's name:	Inês Domingos
Date:	2020
Element category:	Wall
Element type:	Ventilated wall
Fixing system:	At sight
Maintenance target element:	Fiber cement plates 180x100cm
Year of construction:	2020
Reference service life (RSL):	30
Estimate service life (ESL):	34,6
Total facade area [m²]:	219
Replacement cost [€/m²]:	96
Clean cost [€/m²]:	19,49
Inspection cost [€]:	70

Figure 5.
Visualisation of the degree of deterioration and respective values.

it is possible to proceed to the selection of all the elements present in the view 'Analysis of useful life' where the two façades and the roof under analysis are inserted. Next, the 'VUE Analysis' script requires the indication of the year of construction and the year for which the user wants to make the prediction of the state of degradation of the element under analysis. In the lower area of the interface, it is displayed the colour legend that has been assigned according to different intervals of the % lifetime value. The execution of each of the scripts over the case study went smoothly and no errors were reported by Dynamo. After, an Excel table is obtained and exported supporting professionals to define or verify maintenance actions or plans.

This script considers the prediction of the useful life, in a global way. In the second script, an adequate colour is associated to a complete element (façade or terrace), and, as so, it not possible to define distinct zones or areas of an element with different level of deterioration. It allows to make a general comparison between the elements of the building and can even lead to rethinking the choice of some materials so that they are closer to the life of the project. However, in a more realistic situation, a more fragmented modelling of each of the elements could be applied.

6. Results and discussion

In the occupation-building phase, the BIM methodology can be implemented in order to improve the maintenance activity, namely, in the support of retrieving the required information within the database of the model and in the capacity of promoting an agile handling of data used for maintenance propose. The principal aim of the study was to incorporate into the BIM model, which must be created for each building in analyses, the degradation perdition value concerning some of the components of the building, with an important role in the exterior protection of the building, the façades and roofs.

Using the perdition value, a new strategy of visual impact was introduced supporting experts to improve their studies and easily present results to other partners and building owners, as it is possible to observe with distinct colours applied over the model, the level of degradation of the components for a horizontal period of time. In it, two Dynamo scripts were developed.

The application of the scripts developed for the case study allowed to achieve the objectives intended for the study, allowing to make an analysis of the durability of the elements to be applied in the building under study, in the design phase where the solution changes have less impact at the global costs. However, the BIM objects did not have the necessary parameters for the application of the factor method. It needs to include the reference useful life (RUL) found in the digital catalogue of each material, in order to evaluate the durability of the constructive elements in the project to which they are applied, taking into account all the factors to be filled in by the maintenance expert.

However, for the application of these scripts to a BIM model, it is necessary to create the design parameters: ESL, Excel Maintenance, % Useful life and the parameters related to costs, such as mentioned, because that's what they were called when programming was done in Dynamo. At this point, it was necessary to include, in the definition of each script, the creation of these parameters, using the node Parameter. Whenever the script was executed, the parameters were created and in the next execution, there was duplication of parameters. It would be more efficient if these new parameters were added directly and previously in Revit. Alternatively, one could proceed to the programming of independent scripts to create the necessary parameters, but the goal was to be all integrated into the same script.

The export of all the information to Excel was successful, and Dynamo proved to be a good programming software to collect and manage information from the initial model for the maintenance phase such as the composition of the elements, the corresponding area and the associated costs that served as the basis for the proposed maintenance plan. The file is also available from the BIM model, which can be accessed by selecting the Excel Maintenance parameter of the respective element and can be updated throughout its useful life.

7. Conclusion

Programing in Dynamo improved BIM performance concerning the maintenance activity, namely, concerning the estimation of the service live values of the selected materials, frequently applied in façades and terraces as finishing elements. The developed scripts allowed an analysis of the durability of the elements in the design phase where the change of solutions has less impact on overall costs. These scripts, elaborated for the presented study case, can be easily adapted to other components of the building and to other buildings. Two new parameters must be added to parametric objects representative of other components, of the present model, or of other building cases, and then the described procedure can be applied.

The application of the scripts developed for the case study allowed to achieve the objectives intended for the study, allowing to make an analysis of the durability of the elements to be applied in the building under study in the design phase where the change of solutions has less impact on the level of general costs. In it, it was necessary to assign new parameters for the application of the factorial method. Highlighting the need to include the reference useful life (ESL) in the digital catalogue of each object in order to evaluate the durability of the constructive elements in the project to which they are applied, taking into account all the factors to be filled in by the technician. The execution of each of the scripts for the case study went smoothly and no errors were reported by Dynamo.

In addition, an architectural BIM model was developed, where only the essential steps for the modelling of the analysed elements were described, also illustrating the insertion of some types of data and additional parameters in order to enrich the BIM model with relevant information for this analysis. It is also notorious for a computer aspect in the skills of a civil engineer that increasingly tends to be necessary to explore more complex scenarios. This led to the learning of visual programming in Dynamo, and the textual syntax language in Python was also used to achieve the desired objectives through a self-learning process using tutorials available online that allowed us to explore the capabilities of this tool. Since the use of the BIM methodology is increasingly evident and the curricular plan attended does not guarantee appetite in the field of this application, this work allowed to acquire knowledge of its use.

The proposed approach supports the maintenance engineers as decision-makers concerning the maintenance activity. The developed Dynamo scripts improved the BIM-FM activity, providing the maintenance engineer with a prediction of the service life of the materials analysed, and bringing a positive contribution in the context of the building maintenance.

Author details

Alcínia Zita Sampaio*, Inês Domingos and Augusto Gomes
Deparment of Civil Engineering, University of Lisbon, Lisbon, Portugal

*Address all correspondence to: zita.sampaio@tecnico.ulisboa.pt

References

[1] Guzman G, Ulloa W. BIM application in the operation and maintenance management of a sports infrastructure. In: Tommelein ID, Daniel E, editors. Proceedings of the 28th Annual Conference of the International Group for Lean Construction (IGLC28), 6-12 July 2020; Berkeley, CA, USA: International Group for Lean Construction. 2020:949-960. DOI: 10.24928/2020/0015

[2] Sampaio AZ. Maturity of BIM implementation in the construction industry: Governmental policies. International Journal of Engineering Trends and Technology (IJETT). 2021;**69**(7):92-100. Available from: http://www.ijettjournal.org/archive/ijett-v69i7p214

[3] Neuville R, Pouliot J, Billen R. Identification of the best 3D viewpoint within the BIM model: Application to visual tasks related to facility management. Buildings. 2019;**9**:167. DOI: 10.3390/buildings9070167

[4] Fargnoli M, Lleshaj A, Lombardi M, Sciarretta N, Di Gravio G. A BIM-based PSS approach for the management of maintenance operations of building equipment. Buildings. 2019;**9**(6):139. DOI: 10.3390/buildings9060139

[5] Bozorgi PP, Gao X, Eastman C, Self AP. Planning and developing facility management-enabled building information model (FM-enabled BIM). Automation in Construction. 2018;**87**:22-38. DOI: 10.1016/j.autcon.2017.12.004

[6] Sampaio AZ, Simões DG, Berdeja EP. BIM tools used in maintenance of buildings and on conflict detection. In: Sustainable Construction: Building Performance Simulation and Asset and Maintenance Management. Series: Building Pathology and Rehabilitation. Vol. 8. Springer Nature. 2016. pp. 163-183. Available from: http://www.springer.com/gp/book/9789811006500

[7] Domingos I. The 7D/BIM Model Used in the Maintenance of Buildings. Lisbon, Portugal: University of Lisbon; 2020

[8] Sampaio AZ. BIM training course improving skills of construction industry professionals. In: Martinho R, Rijo R, Cunha MM, editors. Procedia Computer Science, special issue: International Conference on Project MANagement (ProjMAN) 2022. Elsevier - ScienceDirect. Vol. 219, 2023, 2035-2042. DOI: 10.1016/j.procs.2023.01.505

[9] Gao X, Pishdad-Bozorgi P. BIM-enabled facilities operation and maintenance: A review. Advanced Engineering Informatics. 2019;**39**:227-247. DOI: 10.1016/j.aei.2019.01.005

[10] Jang R, Collinge W. Improving BIM asset and facilities management processes: A mechanical and electrical (M&E) contractor perspective. Journal of Building Engineering. Elsevier. 2020;**32**:101540. DOI: 10.1016/j.jobe.2020.101540

[11] Ensafi M, Thabet W, Afsari K, Yang E. Challenges and gaps with user-led decision-making for prioritizing maintenance work orders. Journal of Building Engineering. 2023;**66**:105840. DOI: 10.1016/j.jobe.2023.105840

[12] Gouda A, Mohamed M, Abdallah R, Marzoukc M. BIM and semantic web-based maintenance information

for existing buildings. Automation in Construction. 2020;**116**:103209. DOI: 10.1016/j.autcon.2020.103209

[13] Sampaio AZ, Gomes AM. BIM interoperability in structural design, MDPI – Civil engineering, special issue: Advances in civil engineering. Civil Engineering. 2021;**2**(1):174-192. DOI: 10.3390/civileng2010010

[14] Pavón RM, Alberti MG, Álvarez AAA, Carrasco DRC, I. Use of BIM-FM to transform large conventional public buildings into efficient and smart sustainable buildings. Energies. 2021;**14**(11):3127. DOI: 10.3390/en14113127

[15] Hovde PJ, Moser K. Performance based methods for service life prediction. In: State of the Art Reports. Vol. 294. CIB Report. Copyright © 2004 by CIB, NTNU and EMPA, Norway: NTNU – Norwegian University of Science and Technology. 2004. Available from: https://www.researchgate.net/publication/305618294

[16] Sacks R, Eastman C, Lee G, Teicholz P. BIM Handbook: A Guide to Building Information Modeling For Owners, Managers, Architects, Engineers and Contractors. Third ed. New Jersey, USA: John Wiley & Sons, Inc; 2018. DOI: 10.1002/9781119287568

[17] Sampaio AZ, Gomes AM, Lite AS, Zulueta P, González-Gaya C. Analysis of BIM methodology applied to practical cases in the preservation of heritage buildings. MDPI – Sustainability, Special issue: Strategies for the Conservation of Industrial Heritage Aligned with the Sustainable Development Goals. 2021;**13**:3129. DOI: 10.3390/su13063129

[18] Tsay GS, Staub-French S, Poirier É. BIM for facilities management: An investigation into the asset information delivery process and the associated challenges. Applied Science. 2022;**12**:9542. DOI: 10.3390/app12199542

[19] Lavy S, Saxena N, Dixit M. Effects of BIM and COBie database facility management on work order processing times: Case study. Journal of Performance of Constructed Facilities. 2019;**33**(6):04019069

[20] Munir M, Kiviniemi A, Jones S, Finnegan S. BIM-based operational information requirements for asset owners. Architectural Engineering and Design Management. 2020;**16**(2):100-114. DOI: 10.1080/17452007.2019.1706439

[21] Raposo T. Durability of the Building: Estimated Life of Flat Cover Coatings. Porto, Portugal: University of Porto, Faculty of Engineering; 2009

[22] Lopes C, Durability in Construction. Estimated Life of façade Ceramic Coatings. Porto, Portugal: University of Porto, Faculty of Engineering; 2009

[23] Matos M. Durability as a Design Criterion. The Factorial Method in the Portuguese Context. Porto, Portugal: University of Porto, Faculty of Engineering; 2007

[24] ISO 15686-1. Buildings and Constructed Assets - Service Life Planning – Part 1: General Principles. ISO - International Organization for Standardization, Geneva. 2000. Available from: https://www.iso.org/obp/ui/#iso:std:iso:15686:-5:ed-2:v1:en [Accessed June 10, 2023]

[25] Dynamo. The Dynamo Primer. 2020. Available from: https://primer.dynamobim.org [Accessed June 10, 2023]